ISBN: 978-1-7640641-7-0

Published by **Wattage Publishers**
www.WattagePublishers.com
www.HarshaBuilders.com
www.WierCampus.com

PDF and EPUB versions of this book are available at
www.WattagePublishers.com

This page intentionally left blank.

Preface

Designing rectangular concrete columns under biaxial bending remains one of the most intricate challenges in civil and structural engineering. While Eurocode 2 offers a comprehensive design framework, it can still feel abstract or inaccessible in practice.

Deep Surface: Unlocking N-M Biaxial Interaction in Rectangular Concrete Columns to Eurocode 2 was written to bridge that gap. This book presents a clear, structured pathway for generating both 2D interaction diagrams and full 3D N-M surfaces in compliance with Eurocode 2.

Albeit, design software is widely available in modern practice and can perform these calculations in seconds, a true structural or civil engineer must understand what is actually happening inside the column. No software can replace engineering judgement and that judgement comes from a solid grasp of the core theory, assumptions, and behaviour behind the calculations. Even the best software must be created, validated, and used by people who understand the mechanics deeply. This book supports that level of understanding.

At its core, this book is about unlocking a deeper view, one that goes beyond isolated calculations to reveal the full behaviour of columns under complex loading. By generating your own design diagrams and interaction surfaces, you will gain both confidence and clarity in tackling biaxial column challenges.

I hope this book empowers you to explore biaxial behaviour with fresh clarity and unlock your own "Deep Surface" of understanding.

Dr. Harshana S. P. Wattage
Sydney, Australia
May/2025

This page intentionally left blank.

About the Author

Doctor Harshana S. P. Wattage is an experienced university lecturer and researcher with a strong professional background and entrepreneurial skills in the commercial construction industry, along with doctorate-level qualifications in engineering. He holds a PhD in Civil Engineering from an Australian university, with a research concentration in Structural Engineering, and has produced award-winning research publications and patents. His work reflects a strong focus on professional, academic, and research-driven initiatives, alongside a forward-looking interest in cryptocurrency, artificial intelligence, and emerging technologies. Originally from Sri Lanka and now an Australian citizen, he brings a globally informed perspective to both academic and professional domains.

Let's Connect

Facebook: www.facebook.com/harshana1
Instagram: www.instagram.com/xx_harshana_xx
LinkedIn: www.linkedin.com/in/harshanawattage
Google Scholar: scholar.google.com/citations?user=-WmmI0gAAAAJ

This page intentionally left blank.

Contents

Preface iii

About the Author v

1 Introduction and Fundamental Concepts 1

1.1 Introduction 1

1.2 Fundamental Concepts 3

1.2.1 Stress 4

1.2.2 Strain 4

1.2.3 Significance of Stress and Strain 5

1.2.4 Strain at Peak Stress and Ultimate Strain 6

1.2.5 Concrete Strength Classifications 7

1.2.6 Yielding in Materials and its Role in Structural Design 8

1.2.6.1 Steel Yielding in Reinforced Concrete . 9

1.2.6.2 Rationale for Using Yield Strength of Steel in Both Compression and Tension 9

2 Area of Concrete Compression Zone 11

2.1 Calculation of Inclination Angle (θ) of Neutral Axis . . . 11

2.1.1 Method of Calculation 11

2.1.2 Example Calculation 12

2.2 The Lever Arm 'La_{nac}' 14

2.2.1 Method of Calculation 14

2.2.2 Example Calculation 15

2.3 Compression Zone Limits ($\alpha\delta b$ and $\alpha\delta h$) 16

2.3.1 Method of Calculation 16

2.3.2 Example Calculation 19

2.4 Calculation of Area of Concrete in Compression (A_{cc}) . . 19

2.4.1 Behaviour Case 1: $\alpha\delta b \leq b$, and $\alpha\delta h \leq h$ 21

2.4.1.1 Method of Calculation 21

2.4.1.2 Example Calculation for Behaviour Case 1: $\alpha\delta b \leq b$, and $\alpha\delta h \leq h$ 22
2.4.2 Behaviour Case 2: $\alpha\delta b > b$, and $\alpha\delta h \leq h$ 22
2.4.2.1 Method of Calculation 22
2.4.2.2 Example Calculation for Behaviour Case 2: $\alpha\delta b > b$, and $\alpha\delta h \leq h$ 24
2.4.3 Behaviour Case 3: $\alpha\delta b \leq b$, and $\alpha\delta h > h$ 26
2.4.3.1 Method of Calculation 26
2.4.3.2 Example Calculation for Behaviour Case 3: $\alpha\delta b \leq b$, and $\alpha\delta h > h$ 28
2.4.4 Behaviour Case 4: $\alpha\delta b > b$, and $\alpha\delta h > h$ 29
2.4.4.1 Method of Calculation 29
2.4.4.2 Example Calculation for Behaviour Case 4: $\alpha\delta b > b$, and $\alpha\delta h > h$ 32

3 Centroids and Lever Arms for Concrete **35**
3.1 The Lever Arms Lax_{mrc} and Lay_{mrc} 35
3.1.1 Behavior Case 1: $\alpha\delta b \leq b$ and $\alpha\delta h \leq h$ 36
3.1.1.1 Method of Calculation 36
3.1.1.2 Example Calculation for Behaviour Case 1: $\alpha\delta b \leq b$, and $\alpha\delta h \leq h$ 37
3.1.2 Behaviour Case 2: $\alpha\delta b > b$, and $\alpha\delta h \leq h$ 38
3.1.2.1 Method of Calculation 38
3.1.2.2 Example Calculation for Behaviour case 2: $\alpha\delta b > b$, and $\alpha\delta h \leq h$ 40
3.1.3 Behaviour Case 3: $\alpha\delta b \leq b$, and $\alpha\delta h > h$ 41
3.1.3.1 Method of Calculation 41
3.1.3.2 Example Calculation for Behaviour Case 3: $\alpha\delta b \leq b$, and $\alpha\delta h > h$ 43
3.1.4 Behaviour Case 4: $\alpha\delta b > b$, and $\alpha\delta h > h$ 43
3.1.4.1 Method of Calculation 44
3.1.4.2 Example Calculation for Behaviour Case 4: $\alpha\delta b > b$, and $\alpha\delta h > h$ 47
3.2 The Lever Arms Lax_{mrci} *and* Lay_{mrci} 49
3.2.1 Method 1 49
3.2.1.1 Method of Calculation 50

3.2.1.2 Example Calculation 51
3.2.2 Method 2 51
3.2.2.1 Distance from Inclined (Rotated) x Axis (Lax_{mrci}) 52
3.2.2.1.1 Method of Calculation 52
3.2.2.1.2 Example Calculation 54
3.2.2.2 Distance from Inclined (Rotated) y Axis (Lay_{mrci}) 55
3.2.2.2.1 Method of Calculation 56
3.2.2.2.2 Example Calculation 58
3.2.3 Summary of Calculated Distances 60

4 Resistive Axial Force and Moments in Concrete 61
4.1 Strain (ε) and Stress (σ) for the Concrete Compression Zone 61
4.1.1 Method of Calculation 62
4.1.2 Example Calculation 62
4.2 Resistive axial compressive force in concrete (N_c) 63
4.2.1 Method of Calculation 63
4.2.2 Example Calculation 64
4.3 Resistive Bending Moments in Concrete 65
4.3.1 Method 1: Non-inclined (Unrotated) Axis Method 66
4.3.1.1 Method of Calculation 66
4.3.1.2 Example Calculation of Moments (M_{cxn} and M_{cyn}) 68
4.3.1.3 Summary of Calculations 70
4.4 Method 2: Inclined (Rotated) Axis Method 70
4.4.1 Method of Calculation 70
4.4.2 Example Calculation of Moments (M_{cxi} and M_{cyi}) 71
4.4.3 Summary of Calculations 75

5 The Lever Arms for Tensile and Compressive Steel 77
5.1 The Lever Arms Lax_{cfs} and Lay_{cfs} 77
5.2 The Lever Arms Lax_{mrs} and Lay_{mrs} 78
5.2.1 Method of Calculation 78
5.2.2 Example Calculation 80

5.2.3 Summary of Calculations 81
5.3 The Lever Arms Lax_{mrsi} and Lay_{mrsi} 81
5.3.1 Method 1 83
5.3.1.1 Method of Calculation 83
5.3.1.2 Example Calculation 83
5.3.2 Method 2 85
5.3.2.1 Distance from Inclined (Rotated) x Axis (Lax_{mrsi}) 85
5.3.2.1.1 Method of Calculation 85
5.3.2.1.2 Example Calculations 88
5.3.2.2 Distance from Inclined (Rotated) y Axis (Lay_{mrsi}) 91
5.3.2.2.1 Method of Calculation 91
5.3.2.2.2 Example Calculation 94
5.3.2.3 Summary of Calculations 98
5.4 The lever arm La_{nas} 98
5.4.1 Method of Calculation 99
5.4.2 Example Calculation 100

6 Resistive Axial Force and Moments in Steel 103
6.1 Resistive Axial Force in Steel (N_s) 103
6.1.1 Strain in Steel 104
6.1.1.1 Method 1 - Non-inclined (Un-rotated) Axes Method 104
6.1.1.1.1 Method of Calculation 104
6.1.1.1.2 Example Calculations 106
6.1.1.2 Method 2 - Inclined (Rotated) Axis Method 111
6.1.1.2.1 Method of Calculation 111
6.1.1.2.2 Example Calculations 114
6.1.1.3 If the Column is in Pure Compression . 116
6.1.1.3.1 Method of Calculation 116
6.1.1.3.2 Example Calculation 118
6.1.2 Design Stress in Steel (σyd_{bc})(Maximum Allowable Stress in Steel) 118
6.1.2.1 Method of Calculation 118
6.1.2.2 Example Calculation 120

6.1.3 Area of Steel (A_s) 122
6.1.3.1 Method 1: Steel Area for Constructing the N-M Digram Based on Minimum and Maximum Reinforcement Percentages 123
6.1.3.1.1 Method of Calculation 123
6.1.3.1.2 Example Calculation of Maximum Steel Area for Bar in Consideration (As_{Mb}) 123
6.1.3.2 Method 2: Steel Area for Constructing the N-M Digram with Custom Bar Parameters 124
6.1.3.2.1 Method of Calculation 124
6.1.3.2.2 Example Calculation 125
6.1.3.3 Method 3: Normalized Reinforcement Index (I_{nr}) and the Area of Steel (A_s) . 125
6.1.3.3.1 Method of Calculation 125
6.1.3.3.2 Example Calculation 126
6.2 Calculation of Axial Force of Steel (N_s) 128
6.2.1 Method of Calculation 128
6.2.1.1 Example Calculation 129
6.2.1.2 Summary of Calculations 130
6.3 Moment Calculation of Each Axis for Steel 130
6.3.1 Method 1 - Non-inclined (Un-rotated) Axis Method 130
6.3.1.1 Method of Calculation 131
6.3.1.2 Example Calculation of Moments (M_{sxn} and M_{syn}) 132
6.3.2 Method 2 - Inclined (Rotated) Axis Method . . . 135
6.3.2.1 Method of Calculation 135
6.3.2.2 Example Calculation of Moments (M_{sxi} and M_{syi}) 137
6.4 Summation of the Axial Forces and Moments of Steel and Concrete for Each Axis 140
6.4.1 Non-inclined (Unrotated) Axes 140
6.4.1.1 Method of Calculation 140

6.4.1.2 Example Calculation 141
6.4.2 Inclined (Rotated) Axes 142
6.4.2.1 Method of Calculation 142
6.4.2.2 Example Calculation 142

7 Resultant Moment and Direction of Steel and Concrete 145
7.1 Resultant Moment of Steel and Concrete 145
7.1.1 Method of Calculation 146
7.1.2 Example Calculation 148
7.2 Direction of the Resultant Resistive and Demand Moments for Steel and Concrete 149
7.2.1 Non-inclined (Unrotated) Axes 149
7.2.1.1 Method of Calculation 149
7.2.1.2 Example Calculation 151
7.2.2 Inclined (Rotated) Axes 152
7.2.2.1 Method of Calculation 152
7.2.2.2 Example Calculation 154

8 Plotting the Two-dimensional (2D) Interaction Diagrams 157
8.1 Reinforcement Depth Ratio (R_{dr}) 157
8.1.1 Method of Calculation 157
8.1.2 Example Calculation 158
8.2 Applied Moment to Maximum Moment Ratio (R_m) . . . 159
8.2.1 Method of Calculation 159
8.2.2 Example Calculation 160
8.3 Two-dimensional (2D) Interaction Diagrams of Axial Force (N) and Total Moments (M_{tr}) 161
8.4 Normalized Two-dimensional (2D) Interaction Diagrams with Dimensionless Parameters 167

9 Plotting the Three-dimensional (3D) Interaction Surface 171
9.1 Wireframe views of 3D Interaction surface 173
9.2 Rendered views of 3D Interaction Surface 175

Appendix **181**
A - Data of 2D Interaction Diagrams 182
B - Supplementary uniaxial design diagrams 185

This page intentionally left blank.

Chapter 1:

Introduction and Fundamental Concepts

1.1 Introduction

Eurocode standards provide the limits and design guidelines for reinforced concrete columns, ensuring that axial loads, bending moments, and other effects are accounted for within safe and efficient design parameters. These standards define the permissible strength, stability criteria, and interaction limits necessary for structural integrity.

For designs requiring unique reinforcement configurations or custom arrangements, custom N-M (axial force-moment) interaction diagrams or surfaces can be produced. This approach is particularly useful in both standard symmetrical layouts and when the reinforcement layout deviates from standard configurations, namely, asymmetrical arrangements or other non-standard placements, allowing the design to be tailored to specific project or design requirements. This method can accommodate a wide range of reinforcement setups, from the simplest configurations to complex arrangements, including asymmetrical or multi-layered reinforcement layouts. By generating custom interaction diagrams or directly utilizing them, one can accurately capture the behaviour of columns under various load scenarios and reuse these diagrams for similar projects in the future. This method provides a high degree of flexibility and precision, ensuring the design aligns with specific safety requirements and structural demands.

Columns are designed based on their loading behaviour, which can be classified into three main types, axial, uniaxial and biaxial. The Axial loading involves forces applied along the central axis of the column, resulting in uniform compression or tension throughout the column without inducing bending moments. The uniaxial loading applies a combination of axial load and bending about one principal axis, either x or y. This causes bending in a single direction while the column also experiences axial compression or tension. The biaxial loading combines an axial load with bending moments about both principal axes, x and y. This requires a more complex analysis, as the column must resist bending moments in both directions simultaneously. Each loading behaviour impacts the design approach, influencing the choice of methods and calculations necessary to ensure stability and strength.

In addition to those loading behaviours, columns may experience eccentric, torsional, and combined loading in certain structural configurations. Eccentric loading occurs when the axial load is applied off-centre, creating additional bending moments around one or both principal axes. This eccentricity introduces additional stresses and must be considered when loads are not perfectly aligned with the centre of the column. Torsional loading involves twisting forces that cause the column to rotate about its longitudinal axis. Although torsion is typically not a primary design concern for columns, it can arise in asymmetrical or irregular structures under lateral loads such as wind or seismic forces. Eurocode 2 readily addresses torsional effects as part of its comprehensive design guidelines, particularly within normal loading conditions. It provides interaction formulas that account for the combined effects of axial, bending, shear, and torsional stresses without the need for separate calculations unless torsion is unusually high or critical. Combined loading refers to scenarios where a column experiences a mix of axial, bending, shear, and possibly torsional forces simultaneously, which is common in real-world structures. The integrated approach of Eurocode considers these combined effects within standard design checks, provided that the loading conditions remain within normal, expected ranges. This ensures that columns are designed to withstand complex loading scenarios efficiently without the need for isolated torsional calculations, simplifying the design process while maintaining structural integrity.

The methods of designing columns revolve around the calculation of N-M interactions, the relationship between axial load (N) and bending moment (M) in a column. This interaction illustrates how these forces combine to impact the capacity of the column and stability. Columns rarely experience purely axial loads. They typically encounter a mix of vertical loads and bending moments due to lateral forces, asymmetry, eccentric loading, and forces transferred from attached beams. Understanding the N-M interaction is essential for assessing the ability of the column to withstand these combined forces without compromising structural integrity.

To ensure safe design limits and code compliance, the N-M interaction diagrams and surfaces are utilized. This method is suitable for manual calculations, spreadsheets and software applications, which provides a precise assessment of the resistance of the column on axial force and bending moments, and allowing a check to see if the reinforcement setup of the column lies within a safe zone. These help verify that columns can handle the intended loads and bending moments effectively, contributing to a stable and reliable structural design.

Moreover, the orientation of the cross-section defines principal axes, commonly labelled as the x and y axes. In this book, the y axis is considered the long axis of the column, regardless of whether the column is oriented horizontally or vertically in the structure. Accordingly, the x-axis (short side) is treated as the major principal axis due to its higher bending stiffness, while the y-axis (long side) is the minor principal axis. This convention helps maintain consistency when referring to bending about the principal axes.

1.2 Fundamental Concepts

Understanding fundamental concepts, namely, stress, strain, and material behaviour is essential to grasp how reinforced concrete components function under load. These concepts form the foundation to analyse both concrete and steel reinforcement, which help determine structural behaviour and ensure safety and reliability.

1.2.1 Stress

Stress is a measure of how much force is applied to a certain area. In other words, if you pull on or push on a material, the material experiences a force spread out over its cross section. The amount of force divided by the area where the force is applied is called stress, which is the force per unit area. This is shown in the Equation 1.1.

$$\text{Stress}(\sigma) = \frac{\text{Force}}{\text{Area}} \tag{1.1}$$

Examples:

Tension (Pulling)
Imagine pulling on the ends of a rope. The rope resists this pulling force, and the cross section of the rope "feels" the force. What the rope feels in every square millimetre is called tensile stress.

Compression (Pushing)
Imagine pressing both ends of a rubber eraser, between your fingers. The eraser "feels" the force pushing inward. What the eraser feels in every square millimetre is called compressive stress.

1.2.2 Strain

Strain is the ratio of deformation of a material when stress is applied. It represents how much a material stretches or compresses relative to its original length. This is shown in the Equation 1.2.

Example: A rubber band gets stretched (strained) when you pull it (stressed).

$$\varepsilon = \frac{\Delta L}{L_0} \tag{1.2}$$

Where,
ε: Strain (unitless)
ΔL: Change in length (μm, mm, m, etc.)
L_0: Original length (μm, mm, m, etc.)

Strain has no units since it is a ratio of two lengths.

1.2.3 Significance of Stress and Strain

Stress and strain are essential to understand how materials behave under forces. They help determine whether a material can withstand a given load without breaking or deforming excessively.

One key reason we use them is that they allow us to calculate the resistive force a material provides when deformed. This is determined by utilizing the elastic modulus (E) of the material, which represents its stiffness, and its cross-sectional area (A).

Initially, stress is determined using the relationship between strain (ε) and Young's modulus (E). This is shown in the Equation 1.3.

$$\sigma = \varepsilon \cdot E \tag{1.3}$$

Where,
σ: Stress
ε: Strain
E: Young's modulus (i.e.: Stiffness)

Since stress is force per unit area, the total resistive axial force (N) is determined as,

$$N = \sigma \cdot A \tag{1.4}$$

Where,
N: Resistive axial force

σ: Stress
ε: Strain
A: Cross sectional area which the force is applied

Consequently, it is utilized to determine the resistive moment as shown in the Equation 1.5.

$$M = N \cdot La \tag{1.5}$$

Where,
M: Resistive bending moment
N: Resistive axial force
La: Lever arm

1.2.4 Strain at Peak Stress and Ultimate Strain

Strain at peak stress (ε_c) is the axial strain at which concrete reaches its maximum compressive stress (σc_{max}). In Eurocode design models, this maximum compressive stress is limited by the design compressive strength (f_{cd}), which is derived from the characteristic compressive strength (f_{ck}). From this point onward, the material does not provide any additional strength but continues to deform until it reaches the ultimate compressive strain (ε_{cu}).

Ultimate compressive strain (ε_{cu}) is the final strain a material experiences before failure. It represents the maximum deformation the material can undergo before breaking or collapsing.

In Eurocode standards, specific notations, namely, 'ε_{c3}' and 'ε_{cu3}' are utilized for concrete. This notation appears in Eurocode and throughout this book. The subscript 'c' indicates the strain at peak stress and 'cu' indicates ultimate compressive strain, while additional numbers like '3' differentiate between stress-strain models defined in Eurocode 2, for different levels of design accuracy.

When using the parabolic-rectangular stress-strain model, 'ε_{c2}' and 'ε_{cu2}' is utilized. For the bilinear stress-strain model, 'ε_{c3}' and 'ε_{cu3}' is utilized instead. Throughout this book, the bilinear stress-strain model is utilized, thereby, 'ε_{c3}' and 'ε_{cu3}' are applied in all calculations. This is shown in Tables 1.1 and 1.2.

Table 1.1: Strain Values at Peak Stress for Concrete

Strain Type	ε_c up to C50/60	ε_c from C51/61 to C90/105
ε_{c1}	0.0018 to 0.00245	0.00246 to 0.0028
ε_{c2}	0.0020	0.0021 to 0.0026
ε_{c3}	0.00175	0.00176 to 0.0023

Table 1.2: Ultimate Strain Values for Concrete

Strain Type	ε_{cu} up to C50/60	ε_{cu} from C51/61 to C90/105
ε_{cu1}	0.0035	0.0034 to 0.0028
ε_{cu2}	0.0035	0.0034 to 0.0026
ε_{cu3}	0.0035	0.0034 to 0.0026

Where,
'C50/60' represents concrete at 28 days of age, with a characteristic cylinder strength of 50 N/mm^2 (50 MPa) and a characteristic cube strength of 60 N/mm^2 (60 MPa).

These values are specified in Eurocode 2 (EN 1992-1-1), Table 3.1.

1.2.5 Concrete Strength Classifications

Concrete strength classifications, for instance, C50/60 indicate the characteristic compressive strength of concrete as defined in Eurocode 2 (EN 1992-1-1). The first number, 'C50' represents the characteristic compressive strength of a cylinder specimen in N/mm^2 (MPa). The second number '60' represents the characteristic compressive strength of a cube specimen in N/mm^2 (MPa).This is shown in Figure 1.1

Example:
C50/60 denotes, 50 N/mm^2 of cylinder strength and 60 N/mm^2 of cube strength.

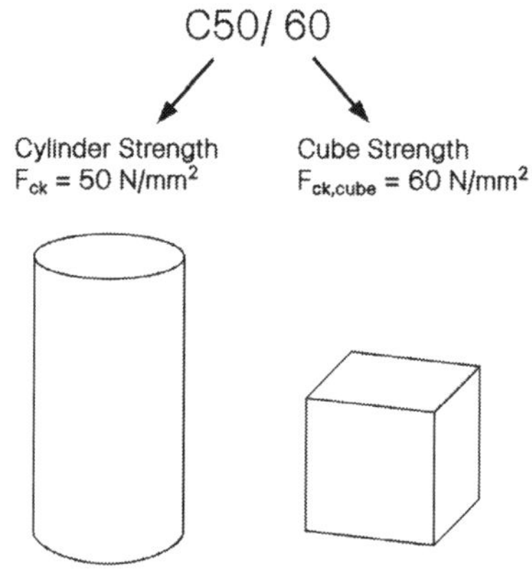

Figure 1.1: Comparison of cylinder and cube strength classification.

1.2.6 Yielding in Materials and its Role in Structural Design

When a material is subjected to force, it first deforms elastically, meaning it returns to its original shape when the force is removed. If the force exceeds a certain limit, known as the yield point, the material enters plastic deformation, where the change in shape becomes permanent.

For instance, a paperclip bent slightly springs back to its original shape (elastic deformation), but if bent too far, it stays deformed (plastic deformation). This transition is called yielding, and it plays a crucial role in structural design, especially in reinforced concrete structures. In these, the embedded steel reinforcement bars primarily undergo longitudinal stretching or shortening, due to axial and bending loads acting on the column. While the reinforcement also follows the curvature of the concrete section and may experience slight bending, this effect is considered structurally negligible. Axial strain is primarily used to analyse the structural behaviour of the reinforcement.

1.2.6.1 Steel Yielding in Reinforced Concrete

To ensure safety, reinforced concrete structures are designed so that steel yields before concrete crushes. Steel is ductile, meaning it can deform significantly before failing, while concrete is brittle and fails suddenly. This behaviour allows steel to act as a warning mechanism, preventing catastrophic failure.

By designing structures so that steel yields before concrete crushes, engineers create ductile failure mechanisms that provide visible warning signs, namely, cracking and deformation, before structural failure. This is why the yield strength (f_y) of steel is used in both tension and compression, simplifying calculations while maintaining safety and reliability.

Since yield strength of steel (f_y) marks the transition between elastic and plastic behaviour, structural calculations focus on the yield strain (ε_y), rather than ultimate strain of steel (ε_{su}). Although, steel does not break at the yield point, this is the critical value used in design to ensure predictable structural performance.

1.2.6.2 Rationale for Using Yield Strength of Steel in Both Compression and Tension

In reinforced concrete design, steel reinforcement is assigned the same yield strength (f_y) for both tension and compression, unlike concrete, which has distinct tensile and compressive strengths. This approach reflects the fundamental behaviour of steel under load.

Concrete has a defined compressive strength, but its tensile strength is often ignored due to its insignificant contribution to tensile resistance. Steel, however, does not have a well-defined compressive strength independent of its tensile yield strength. Instead, its ability to resist compression is directly derived from its tensile yield strength.

This is not a simplification but a necessity, as steel does not fail in compression the way concrete does. Concrete crushes at its ultimate strain, whereas steel undergoes significant plastic deformation in compression without a sudden rupture. Since there is no clear compressive failure mechanism for steel, it is impractical to define a separate compressive strength. Instead, design standards universally apply the tensile yield strength (f_y) for both tensile and compressive reinforcement.

Chapter 2:

Area of Concrete Compression Zone

The area of the concrete compression zone in a biaxial concrete column plays a crucial role in determining the strength and stability of the column. This zone resists compressive forces and directly influences load-bearing capacity, stiffness, and failure mechanisms. Its shape and size vary with axial load and biaxial bending moments, impacting the neutral axis position and moment capacity.

2.1 Calculation of Inclination Angle (θ) of Neutral Axis

The angle of inclination of the neutral axis can be determined by utilizing the arctangent function. This is shown in Equations 2.1 and 2.2.

2.1.1 Method of Calculation

The calculation method of the inclination angle (θ) is shown in Equations 2.1 and 2.2. Moreover this is illustrated in Figure 2.1.

$$IR = \frac{\delta h}{\delta b} \tag{2.1}$$

$$\theta = arctan \cdot (IR) \tag{2.2}$$

Where,

δh: Neutral axis distance along the height.

δb: Neutral axis distance along the breadth.

IR: Inclination ratio.

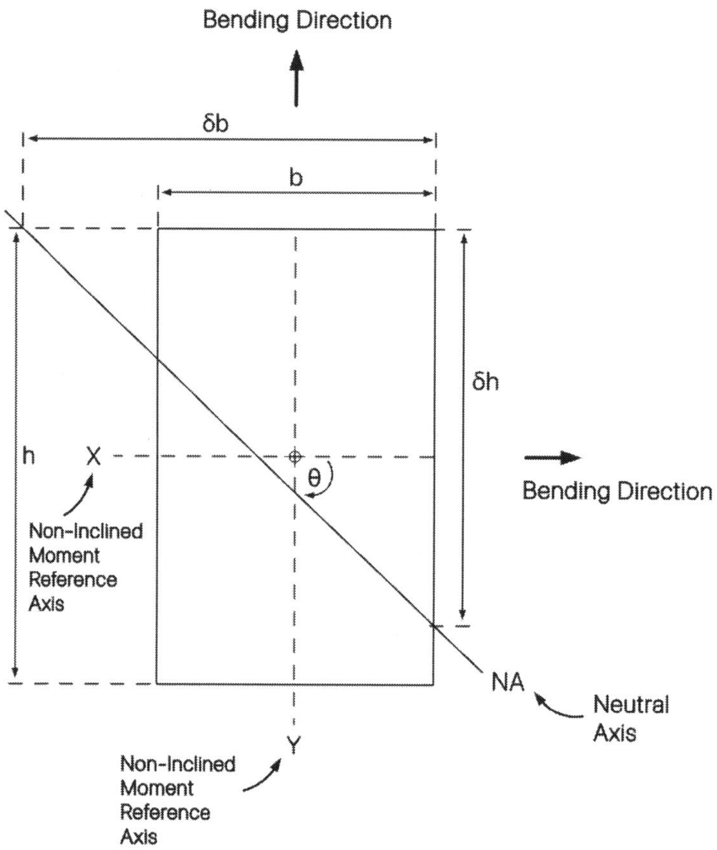

Figure 2.1: Inclination Angle (θ)

θ: Inclination angle (rotation angle) of the neutral axis, relative to the non-inclined (unrotated) x axis.

Note: When using spreadsheet software, ensure that the output of the $arctan$ function is converted to degrees by using $DEGREES(ATAN(IR))$, as some software returns $ATAN$ values in $radians$ by default.

2.1.2 Example Calculation

Given Data:: $\delta h = 200$ mm, $\delta b = 240$ mm.

Calculation of the Inclination Ratio (IR)

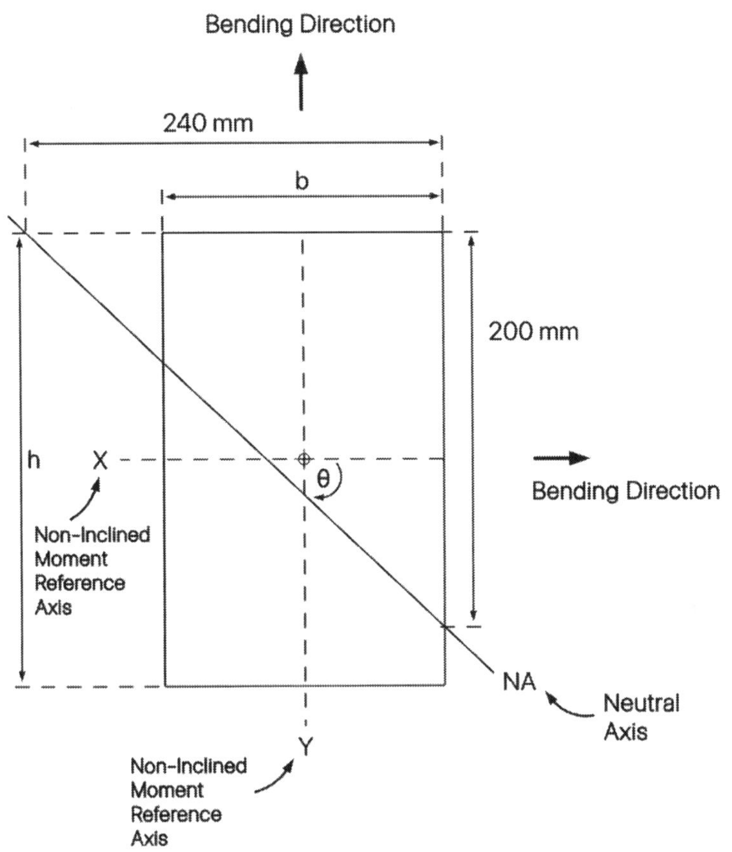

Figure 2.2: Example Calculation of Inclination Angle (θ)

$$\begin{aligned} \text{IR} &= \frac{\delta h}{\delta b} \\ &= \frac{200\text{ mm}}{240\text{ mm}} \\ &= \underline{0.833333333} \end{aligned}$$

Calculation of the Angle of Inclination (θ)

$$\begin{aligned} \theta &= \arctan(\text{IR}) \\ \theta &= \arctan(0.833333333) \\ \theta &= 39.805571092^\circ \end{aligned}$$

2.2 The Lever Arm 'La_{nac}'

This section focuses on the lever arm (La_{nac}), which represents the perpendicular distance from the neutral axis (NA) to the line passing through the point of maximum compressive strain at the extreme compression fibre intersection point (ECFIP).

The extreme compression fibre is the region where the section experiences the highest compressive strain under bending. The extreme compression fibre intersection point (ECFIP) refers to the location where the extreme compression fibres in both the x and y directions meet. By defining this lever arm as a perpendicular distance, it ensures the calculation aligns with the true geometric and strain distribution of the section. This parameter is utilized to determine strain gradients, reinforcement bar strains, and the corresponding stresses and moments of the column. This is shown in the Figure 2.3.

2.2.1 Method of Calculation

The lever arm La_{nac}, can be calculated utilizing basic trigonometry. This is shown in the Equations 2.3 to 2.5.

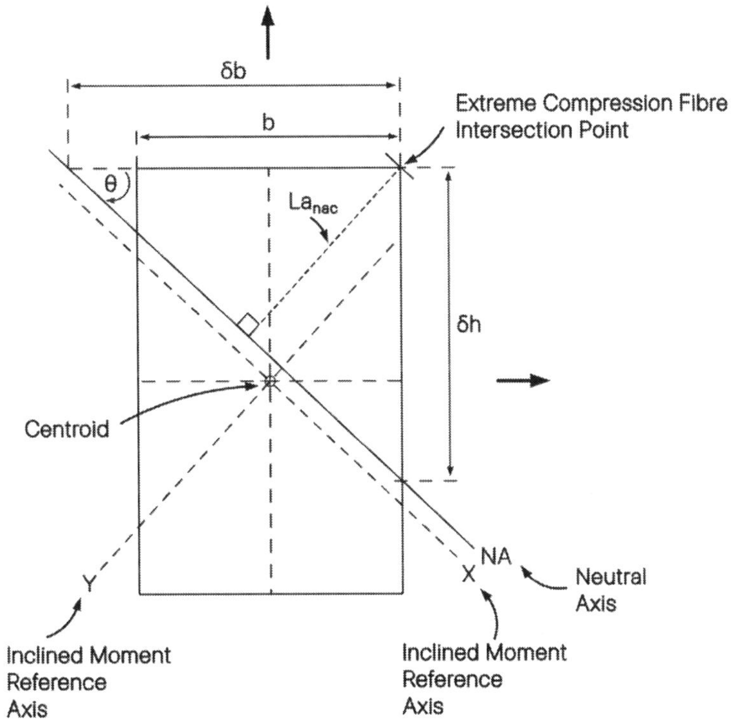

Figure 2.3: The Lever Arm 'La_{nac}'

$$\sin(\theta) = \frac{\text{opposite}}{\text{hypotenuse}} \tag{2.3}$$

$$= \frac{\text{La}_{\text{nac}}}{\delta b} \tag{2.4}$$

$$\text{La}_{\text{nac}} = \sin(\theta) \cdot \delta b \tag{2.5}$$

2.2.2 Example Calculation

Given Data: $\delta h = 200\,\text{mm}$, $\delta b = 240\,\text{mm}$, $\theta = 39.80557109°$

Calculation of $\mathbf{La_{nac}}$:

$$\sin(\theta) = \frac{\text{La}_{\text{nac}}}{\delta b}$$

$$\mathrm{La}_{\mathrm{nac}} = \sin(\theta) \cdot \delta b$$

$$\sin(39.80557109^\circ) = 0.640184399$$

$$\mathrm{La}_{\mathrm{nac}} = 0.640184399 \cdot 240\,\mathrm{mm}$$

$$\mathrm{La}_{\mathrm{nac}} = \underline{153.644\,\mathrm{mm.}}$$

2.3 Compression Zone Limits ($\alpha\delta b$ and $\alpha\delta h$)

The compression zone of a column, follows a shape similar to a parabolic shape. However, accurately determining this exact parabolic profile in biaxial bending involves complex calculations. To address this challenge, Eurocode 2 (EN 1992-1-1), in Section 3.1.7(3), provides a simplified yet safe approximation by enclosing the parabolic shape within clearly defined boundaries.

In this approach, the compression zone is constrained by two key limits, the compression zone limit along the breadth ($\alpha\delta b$) and the compression zone limit along the height ($\alpha\delta h$). These limit conditions define an imaginary compression zone boundary that runs parallel to the neutral axis (NA).

This imaginary compression zone boundary is positioned at 0.8 ($\lambda = 0.8$) times the distance from the intersection of the extreme compression fibre (La_{nac}) to the neutral axis (NA). The factor of 0.8 for λ is explicitly specified in Eurocode 2 (EN 1992-1-1), clause 3.1.7(3).

2.3.1 Method of Calculation

The calculation of the compression zone limit ($\alpha\delta b$) and ($\alpha\delta h$) is shown in Equations 2.6 to 2.10. Moreover, the parabolic non-simplified compression zone boundary is illustrated in the Figure 2.4, and the simplified compression zone boundary is illustrated in Figure 2.5. The Delta (δ), Alpha (α) and Lambda (λ) are Greek letters utilized for notation.

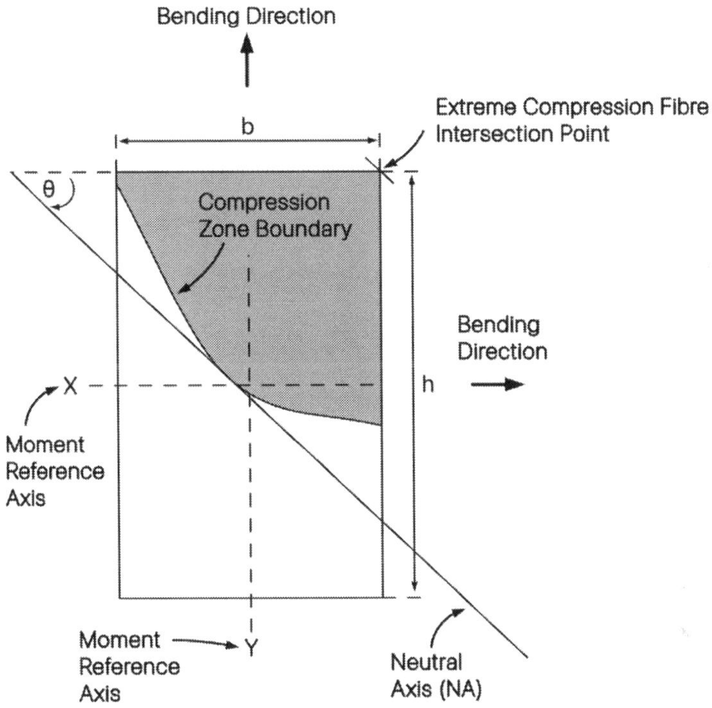

Figure 2.4: Parabolic non-simplified compression zone in biaxial bending.

$$\alpha\delta h = \frac{\lambda \cdot La_{\text{nac}}}{\cos\theta} \tag{2.6}$$

$$\alpha\delta b = \frac{\lambda \cdot La_{\text{nac}}}{\cos(90^\circ - \theta)} \tag{2.7}$$

$$\theta = \arctan(\text{IR}) \tag{2.8}$$

$$\text{IR} = \frac{\delta h}{\delta b} \tag{2.9}$$

$$\lambda = 0.8 \tag{2.10}$$

Where,

$\alpha\delta h$ = Compression zone limit along the height (h)

$\alpha\delta b$ = Compression zone limit along the breadth (b)

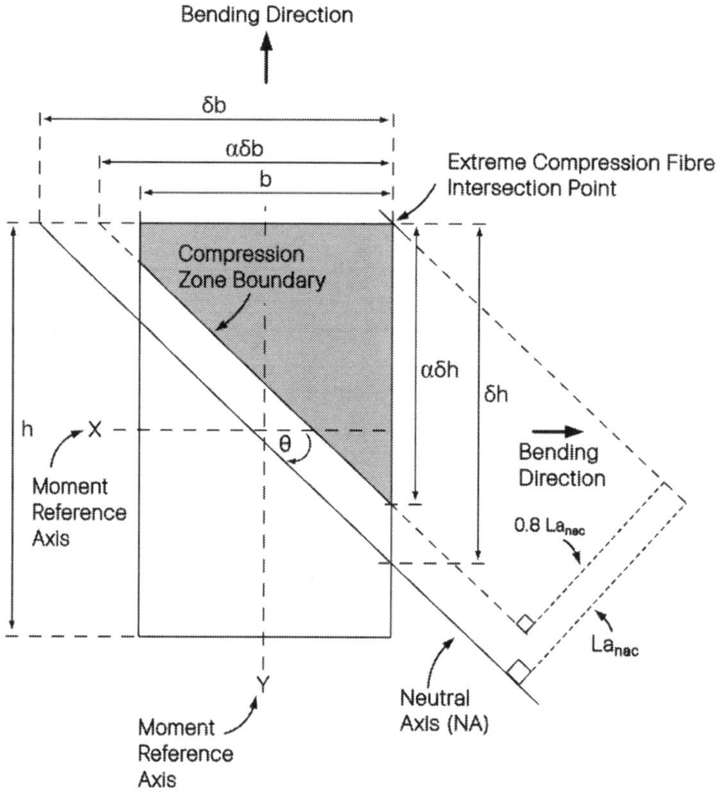

Figure 2.5: Simplified compression zone in biaxial bending.

δh: Neutral axis distance along the height.
δb: Neutral axis distance along the breadth.
IR: Inclination ratio of neutral axis.
θ: Inclination angle (rotation angle) of the neutral axis relative to the non-inclined (unrotated) x axis.
λ: Compression zone reduction factor.
La_{nac}: Distance from the neutral axis (NA) to the extreme compression fibre intersection point (ECFIP).

Note: When using spreadsheet software, ensure that the input for $cos(90 - \theta)$ is written as 'COS (RADIANS (90 - θ))' to convert degrees to radians before applying the cosine function. Some software requires this conversion to interpret the angle correctly.

2.3.2 Example Calculation

Given Data:

$\delta h = 200$ mm, $\delta b = 240$ mm, $\theta = 39.805571092°$, $La_{\text{nac}} = 153.644$ mm, $\lambda = 0.8$.

Calculation of $\alpha\delta h$

$$\begin{aligned}\alpha\delta h &= \frac{\lambda \cdot La_{\text{nac}}}{\cos\theta} \\ &= \frac{0.8 \times 153.644 \text{ mm}}{\cos(39.805571092°)} \\ &= \frac{122.915 \text{ mm}}{0.768221280} \\ &= 160 \text{ mm}\end{aligned}$$

Calculation of $\alpha\delta b$

$$\begin{aligned}\alpha\delta b &= \frac{\lambda \cdot La_{\text{nac}}}{\cos(90° - \theta)} \\ &= \frac{0.8 \times 153.644 \text{ mm}}{\cos(90° - 39.805571092°)} \\ &= \frac{122.915 \text{ mm}}{\cos(50.19442891°)} \\ &= 192 \text{ mm}\end{aligned}$$

2.4 Calculation of Area of Concrete in Compression (A_{cc})

The area of concrete in compression depends on the column behaviour case. These are classified into four main possible compression cases under biaxial loads. Moreover, since there are loads applied in both the x

and y directions, bending occurs around both axes, leading to a biaxial bending scenario.

These cases are illustrated in Figure 2.6, and the Table 2.1 shows the column height (h) and breadth (b) dimensions utilized throughout this section. Moreover, the compression zone limits ($\alpha\delta h$ and $\alpha\delta b$) are provided in table 2.2.

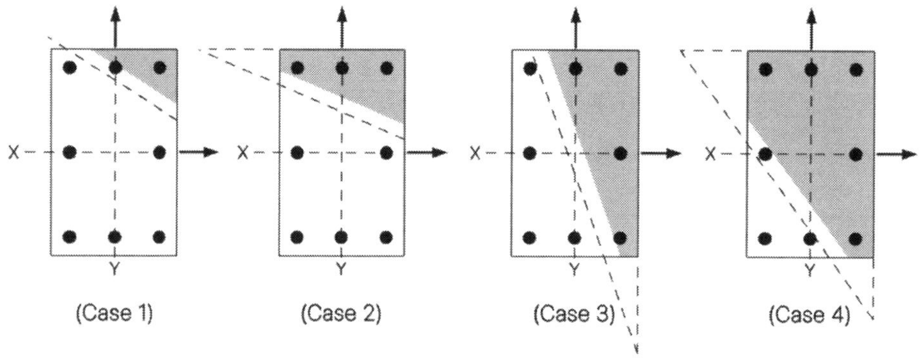

Figure 2.6: Column compression behaviour cases.

Table 2.1: Column Dimensions

Column height (h)	300 mm
Column breadth (b)	225 mm

Table 2.2: Behaviour Case Values

Behaviour Case	$\alpha\delta h$	$\alpha\delta b$
1	160 mm	192 mm
2	75 mm	340 mm
3	375 mm	170 mm
4	375 mm	340 mm

2.4.1 Behaviour Case 1: $\alpha\delta b \leq b$, and $\alpha\delta h \leq h$

The following sections outline the step-by-step approach for determining the compression zone area for the behaviour case 1, including a detailed calculation method and a worked example to illustrate each calculation step in detail.

2.4.1.1 Method of Calculation

This column behaviour occurs when the compression zone limit along the breadth ($\alpha\delta b$) is less than or equal to the column breadth (b), and the compression zone limit along the height ($\alpha\delta h$) is less than or equal to the cross-sectional height (h) of the column.

Since the compression zone in this case is a triangle, the area can be easily determined. This is shown in the Equation 2.11 and depicted in Figure 2.7.

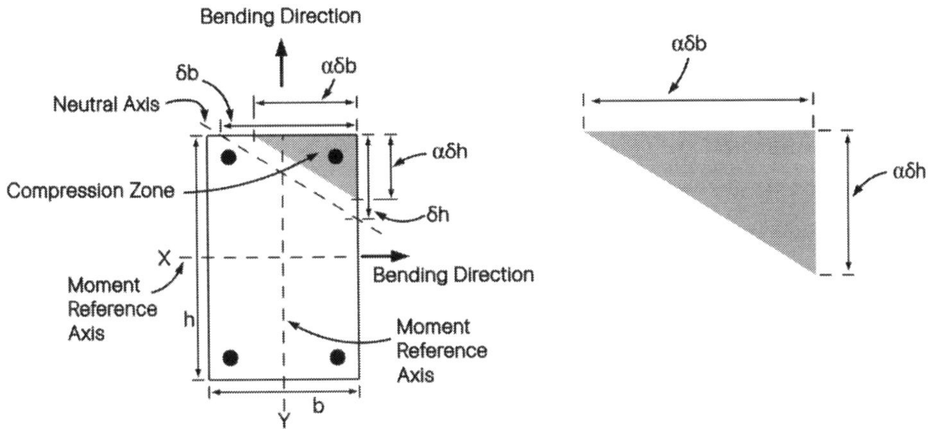

Figure 2.7: Area of compression zone for behaviour case 1

$$A_{cc} = 0.5 \times \alpha\delta b \times \alpha\delta h \tag{2.11}$$

Where,

$\alpha\delta b$: Compression zone limit along the breadth.

$\alpha\delta h$: Compression zone limit along the height.
b: Breadth of the column cross-section.
h: Height of the column cross-section.
A_{cc}: Area of the compression zone of concrete.

2.4.1.2 Example Calculation for Behaviour Case 1: $\alpha\delta b \leq b$, and $\alpha\delta h \leq h$

Given Data: $\alpha\delta h = 160$ mm, $\alpha\delta b = 192$ mm.

The calculation of A_{cc} is as follows.

$$A_{cc} = 0.5 \times \alpha\delta h \times \alpha\delta b$$

$$= 0.5 \times 160 \text{ mm} \times 192 \text{ mm}$$

$$= 15360 \text{ mm}^2$$

2.4.2 Behaviour Case 2: $\alpha\delta b > b$, and $\alpha\delta h \leq h$

The following sections outline the step-by-step approach for determining the compression zone area for the behaviour case 2, including a detailed calculation method and a worked example to illustrate each calculation step in detail.

2.4.2.1 Method of Calculation

This column behaviour occurs when the compression zone limit along the breadth ($\alpha\delta b$) exceeds the column breadth (b), while the compression zone limit along the height ($\alpha\delta h$) remains less than or equal to the cross-sectional height (h) of the column.

The area of the compression zone forms a trapezium. It is determined by first obtaining the necessary data using Equations 2.14 and 2.15, followed by applying the standard trapezium area Equation 2.12. The simplified equation is shown in Equation 2.16 and this is illustrated in the Figure 2.8.

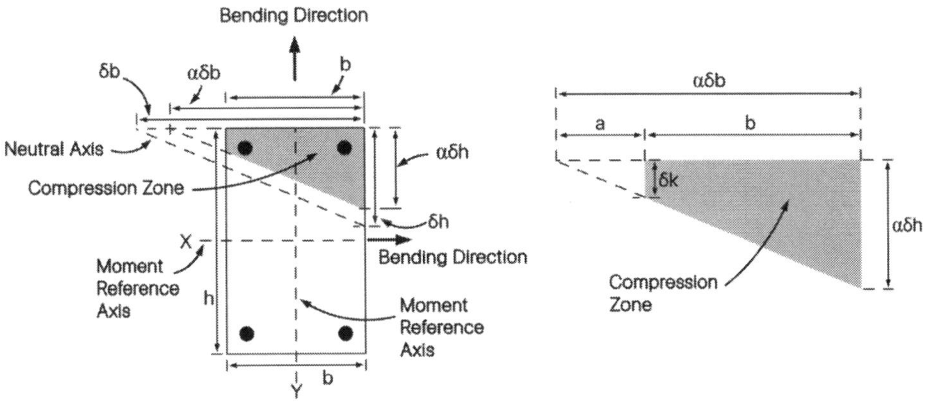

Figure 2.8: Area of compression zone for behaviour case 2

$$\text{Area of trapezium} = \left(\frac{\text{Sum of parallel sides}}{2}\right) \times \text{Height} \tag{2.12}$$

In a similar manner,

$$A_{cc} = \left(\frac{\delta k + \alpha\delta h}{2}\right) \times b \tag{2.13}$$

$$\delta k = \left(\frac{\alpha\delta h}{\alpha\delta b}\right) \times a \leq h \tag{2.14}$$

$$a = \alpha\delta b - b \tag{2.15}$$

Where,

A_{cc} = Area of the compression zone of concrete.
δk = Length of the short side of compression area along the height.
a = The length beyond the actual compression zone along the breadth.
$\alpha\delta b$ = Compression zone limit along the breadth.

$\alpha\delta h$ = Compression zone limit along the height.
b = Breadth of the column cross-section.
h = Height of the column cross-section.

By combining and simplifying all the equations, the Equation 2.16 is derived, which provides a more direct method for calculating the area of the compression zone. This equation remains valid while $\delta\mathrm{k} \leq h$.

Simplified Equation

$$A_{cc} = \frac{\alpha\delta h \cdot b}{2} \cdot \left(2 - \frac{b}{\alpha\delta b}\right) \quad \text{for } \delta k \leq h \tag{2.16}$$

Where,
A_{cc} = Area of the compression zone of concrete.
δk = Length of the short side of compression area along the height.
$\alpha\delta b$ = Compression zone limit along the breadth.
$\alpha\delta h$ = Compression zone limit along the height.
b = Breadth of the column cross-section.
h = Height of the column cross-section.

2.4.2.2 Example Calculation for Behaviour Case 2: $\alpha\delta b > b$, and $\alpha\delta h \leq h$

Given Data: $b = 225$ mm, $\alpha\delta h = 75$ mm, $\alpha\delta b = 340$ mm.

Calculation:

Method 1

$$\begin{aligned} a &= \alpha\delta b - b \\ &= 340 \text{ mm} - 225 \text{ mm} \\ &= 115.000 \text{ mm} \end{aligned}$$

$$\delta_k = \left(\frac{\alpha\delta h}{\alpha\delta b}\right) \cdot a$$

$$= \left(\frac{75 \text{ mm}}{340 \text{ mm}}\right) \times 115.000 \text{ mm}$$

$$= 25.3676470588 \text{ mm}$$

$$A_{cc} = \left(\frac{\delta_k + \alpha\delta h}{2}\right) \cdot b$$

$$= \left(\frac{25.3676470588 \text{ mm} + 75 \text{ mm}}{2}\right) \times 225 \text{ mm}$$

$$= \left(\frac{100.3676470588 \text{ mm}}{2}\right) \times 225 \text{ mm}$$

$$= 50.1838235294 \text{ mm} \times 225 \text{ mm}$$

$$= 11,291.360 \text{ mm}^2$$

Method 2

$$A_{cc} = \frac{\alpha\delta h \cdot b}{2} \cdot \left(2 - \frac{b}{\alpha\delta b}\right)$$

First term:

$$\frac{\alpha\delta h \cdot b}{2} = \frac{75 \text{ mm} \times 225 \text{ mm}}{2}$$

$$= 8,437.500 \text{ mm}^2$$

Second term:

$$\frac{b}{\alpha\delta b} = \frac{225 \text{ mm}}{340 \text{ mm}}$$

$$= 0.6617647059$$

$$2 - \frac{b}{\alpha\delta b} = 2 - 0.6617647059$$

$$= 1.3382352941$$

Combined result for A_{cc}:

$$A_{cc} = 8,437.500 \text{ mm}^2 \times 1.3382352941$$

$$= 11,291.360 \text{ mm}^2$$

$$A_{cc} = \underline{11,291.360 \text{ mm}^2}$$

2.4.3 Behaviour Case 3: $\alpha\delta b \leq b$, and $\alpha\delta h > h$

The following sections outline the approach to determine the compression zone area for the behaviour case 3, including a detailed calculation method and a worked example to illustrate each calculation step in detail.

2.4.3.1 Method of Calculation

This case describes a column behaviour where the compression zone limit along the breadth ($\alpha\delta b$) is less than or equal to the column breadth (b) while the compression zone limit along the height ($\alpha\delta h$) exceeds the cross-sectional height of the column (h).

The maximum possible length of the short side of compression area along the breadth (δv) is equal to the breadth (b) of the column.

The area of the compression zone forms a trapezium. It is determined by first obtaining the necessary data using Equation 2.19, followed by applying the standard trapezium area Equations 2.17 and 2.18. The simplified equation is shown in Equation 2.20 and illustrated in the Figure 2.9.

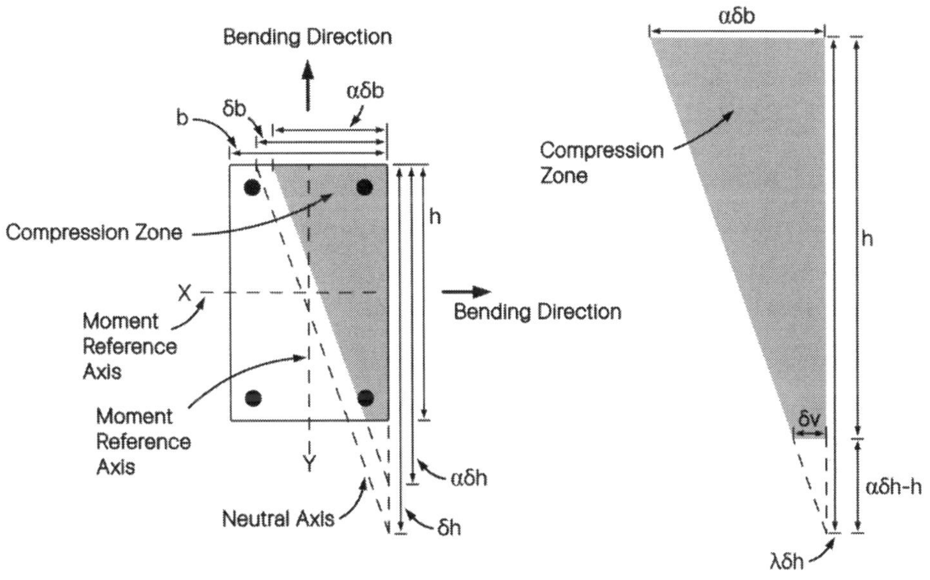

Figure 2.9: Area of compression zone for behaviour case 3

$$\text{Area of trapezium} = \left(\frac{\text{Sum of parallel sides}}{2}\right) \cdot \text{Height} \tag{2.17}$$

In a similar manner,

$$A_{cc} = \left(\frac{\delta v + \alpha\delta b}{2}\right) \cdot h \tag{2.18}$$

$$\delta v = \left(\frac{\alpha\delta b}{\alpha\delta h}\right) \cdot (\alpha\delta h - h) \tag{2.19}$$

Where,

A_{cc} = Area of the compression zone of concrete.
δv = Length of the short side of compression area along the breadth.
$\alpha\delta b$ = Compression zone limit along the breadth.
$\alpha\delta h$ = Compression zone limit along the height.
b = Breadth of the column cross-section.

h = Height of the column cross-section.

Simplified Equation

$$A_{cc} = \frac{\alpha\delta b \cdot h}{2} \cdot \left(2 - \frac{h}{\alpha\delta h}\right) \quad \text{for } \delta v \leq b \tag{2.20}$$

Where,

A_{cc} = Area of the compression zone of concrete.
δv = Length of the short side of compression area along the breadth.
$\alpha\delta b$ = Compression zone limit along the breadth.
$\alpha\delta h$ = Compression zone limit along the height.
b = Breadth of the column cross-section.
h = Height of the column cross-section.

2.4.3.2 Example Calculation for Behaviour Case 3: $\alpha\delta b \leq b$, and $\alpha\delta h > h$

Given Data: $h = 300$ mm, $\alpha\delta h = 375$ mm, $\alpha\delta b = 170$ mm.

Calculation:

Method 1:

$$\begin{aligned}
\delta_v &= \left(\frac{\alpha\delta b}{\alpha\delta h}\right) \cdot (\alpha\delta h - h) \\
&= \left(\frac{170 \text{ mm}}{375 \text{ mm}}\right) \times (375 \text{ mm} - 300 \text{ mm}) \\
&= 34 \text{ mm} \\
A_{cc} &= \left(\frac{\delta_v + \alpha\delta b}{2}\right) \cdot h \\
&= \left(\frac{34 \text{ mm} + 170 \text{ mm}}{2}\right) \times 300 \text{ mm} \\
&= \left(\frac{204 \text{ mm}}{2}\right) \times 300 \text{ mm} \\
&= 102 \text{ mm} \times 300 \text{ mm}
\end{aligned}$$

$$= 30,600.000 \text{ mm}^2$$

Method 2:

$$\begin{aligned} A_{cc} &= \frac{\alpha\delta b \cdot h}{2} \cdot \left(2 - \frac{h}{\alpha\delta h}\right) \\ &= \frac{170 \text{ mm} \times 300 \text{ mm}}{2} \times \left(2 - \frac{300 \text{ mm}}{375 \text{ mm}}\right) \\ &= \frac{51,000 \text{ mm}^2}{2} \times (2 - 0.8) \\ &= 25,500 \text{ mm}^2 \times 1.2 \\ &= 30,600.000 \text{ mm}^2 \end{aligned}$$

2.4.4 Behaviour Case 4: $\alpha\delta b > b$, and $\alpha\delta h > h$

The following sections outline the step-by-step approach for determining the compression zone area for the behaviour case 4, including a detailed calculation method and a worked example to illustrate each calculation step in detail.

2.4.4.1 Method of Calculation

This column behaviour occurs when the compression zone limit along the breadth ($\alpha\delta b$) exceeds the column breadth (b), and the compression zone limit along the height ($\alpha\delta h$) exceeds the cross-sectional height (h) of the column.

In Case 4, the compression zone is determined by subtracting the triangular non-compression zone from the total cross-sectional area of the column. This approach is outlined in Equations 2.22 to 2.24, with the simplified equation shown in Equation 2.25.

The maximum possible length of the long side of the compression area along the height (δk) is equal to the column height (h), while the maximum possible length of the short side of the compression area along the breadth (δv) is equal to the column breadth (b). This is illustrated in the Figures 2.10 and 2.11.

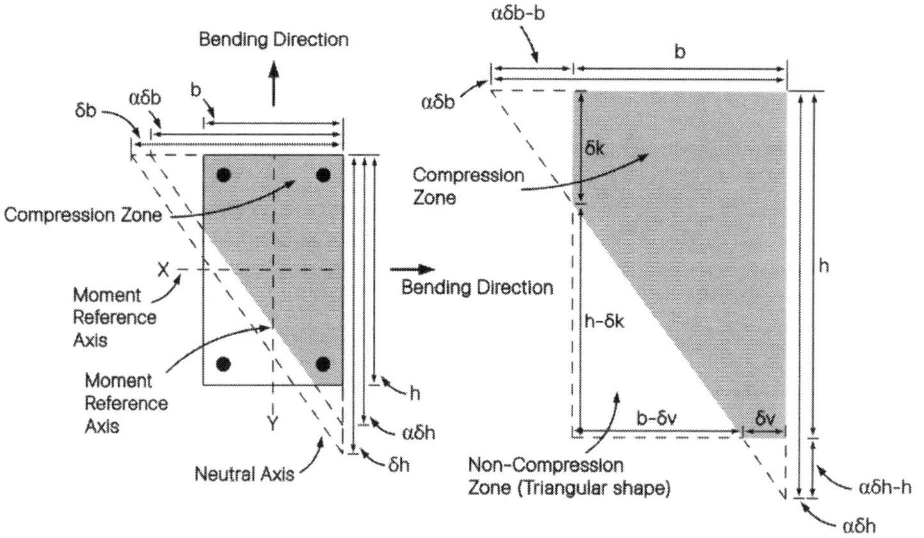

Figure 2.10: Area of compression zone for behaviour case 4

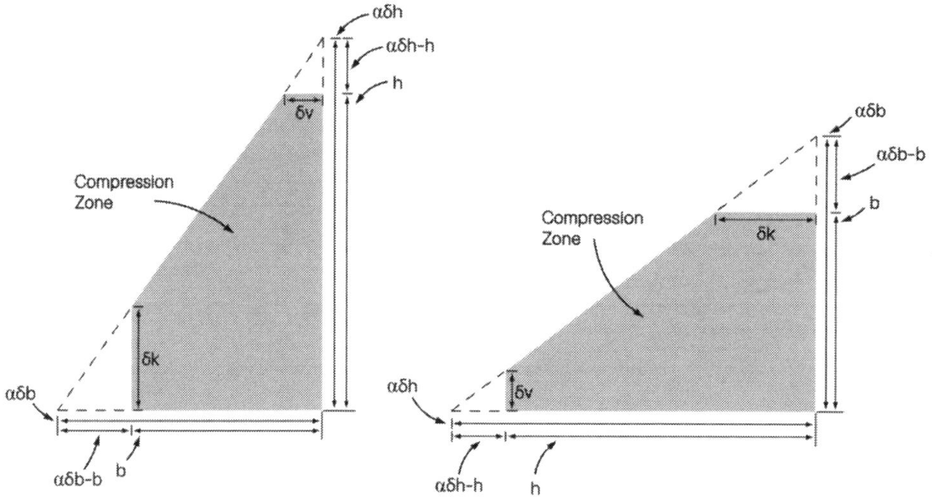

Figure 2.11: Rotated/flipped compression zones aligned to base for simplified calculation of δk and δv

$$A_{Pent} = (b \cdot h) - (A_{Tri}) \tag{2.21}$$

Where,

A_{Pent}: Area of the pentagonal compression zone.
$b \cdot h$: Column cross-sectional area.
A_{Tri}: The area of the triangular non-compression zone.

In a similar manner,

$$A_{cc} = (h \cdot b) - (0.5 \cdot (h - \delta k) \cdot (b - \delta v)) \tag{2.22}$$

$$\delta k = \left(\frac{\alpha\delta h}{\alpha\delta b}\right) \cdot (\alpha\delta b - b), \quad \delta k \leq h \tag{2.23}$$

$$\delta v = \left(\frac{\alpha\delta b}{\alpha\delta h}\right) \cdot (\alpha\delta h - h), \quad \delta v \leq b \tag{2.24}$$

Where,

A_{cc} = Area of the compression zone of concrete.
δv = Length of the short side of compression area along the breadth.
δk = Length of the long side of compression area along the height.
$\alpha\delta b$ = Compression zone limit along the breadth.
$\alpha\delta h$ = Compression zone limit along the height.
$\alpha\delta b - b$ = The length beyond the actual compression zone along the breadth (b).
$\alpha\delta h - h$ = The length beyond the actual compression zone along the height (h).
b = Breadth of the column cross-section.
h = Height of the column cross-section.

Combining and simplifying the above equations result in the following streamlined formula shown in Equation 2.25, providing an efficient method for calculating the area of the compression zone. This equation is valid while $\delta k \leq h$ and $\delta v \leq b$.

Simplified Equation

$$A_{cc} = h \cdot b - 0.5 \cdot \left(\left(h - \left(\frac{\alpha\delta h}{\alpha\delta b} \cdot (\alpha\delta b - b)\right)\right) \cdot \left(b - \left(\frac{\alpha\delta b}{\alpha\delta h} \cdot (\alpha\delta h - h)\right)\right)\right) \tag{2.25}$$

Where,

A_{cc} = Area of the compression zone of concrete.
$\alpha\delta h$ = Compression zone limit along the height.
$\alpha\delta b$ = Compression zone limit along the breadth.
b = Breadth of the column cross-section.
h = Height of the column cross-section.

2.4.4.2 Example Calculation for Behaviour Case 4: $\alpha\delta b > b$, and $\alpha\delta h > h$

Given Data: h = 300 mm, b = 225 mm, $\alpha\delta h$ = 375 mm, $\alpha\delta b$ = 340 mm.

Calculation:

Method 1:

$$\delta_k = \left(\frac{\alpha\delta h}{\alpha\delta b}\right) \cdot (\alpha\delta b - b)$$

$$= \left(\frac{375}{340 \text{ mm}}\right) \times (340 \text{ mm} - 225 \text{ mm})$$

$$= 1.10294 \times 115 \text{ mm} = 126.838 \text{ mm}$$

$$\delta v = \left(\frac{\alpha\delta b}{\alpha\delta h}\right) \cdot (\alpha\delta h - h)$$

$$= \left(\frac{340 \text{ mm}}{375 \text{ mm}}\right) \times (375 \text{ mm} - 300 \text{ mm})$$

$$= 0.90667 \times 75 \text{ mm} = 68 \text{ mm}$$

$$A_{cc} = (h \cdot b) - (0.5 \cdot (h - \delta k) \cdot (b - \delta v))$$

$$= (300 \text{ mm} \times 225 \text{ mm}) - (0.5 \times (300 \text{ mm} - 126.838 \text{ mm}) \times (225 \text{ mm} - 68 \text{ mm}))$$

$$= 67,500 \text{ mm}^2 - (0.5 \times 173.162 \text{ mm} \times 157 \text{ mm})$$

$$= 67,500 \text{ mm}^2 - (0.5 \times 27,186.434 \text{ mm}^2)$$

$$= 67,500 \text{ mm}^2 - 13,593.217 \text{ mm}^2 = 53,906.783 \text{ mm}^2$$

Method 2:

$$A_{cc} = h \cdot b - 0.5 \cdot \left(\left(h - \left(\frac{\alpha\delta h}{\alpha\delta b} \cdot (\alpha\delta b - b) \right) \right) \cdot \left(b - \left(\frac{\alpha\delta b}{\alpha\delta h} \cdot (\alpha\delta h - h) \right) \right) \right)$$

First term:

$$300 \text{ mm} - \left(\frac{375 \text{ mm}}{340 \text{ mm}} \times 115 \text{ mm} \right) = 300 \text{ mm} - 126.838 \text{ mm} = 173.162 \text{ mm}$$

Second term:

$$225 \text{ mm} - \left(\frac{340 \text{ mm}}{375 \text{ mm}} \times 75 \text{ mm} \right) = 225 \text{ mm} - 68 \text{ mm} = 157 \text{ mm}$$

Combined result for A_{cc}:

$$A_{cc} = 67,500 \text{ mm}^2 - 0.5 \cdot (173.162 \text{ mm} \times 157 \text{ mm})$$

$$= 67,500 \text{ mm}^2 - 0.5 \times 27,186.434 \text{ mm}$$

$$= 67,500 \text{ mm}^2 - 13,593.217 \text{ mm}^2 = 53,906.783 \text{ mm}^2$$

This page intentionally left blank.

Chapter 3:

Centroids and Lever Arms for Concrete

Centroids of the concrete compression zone and their corresponding lever arms play a fundamental role in determining the resistive moments of the column.

The centroid, often referred to as the centre of gravity, is the point where the entire area of the compressive zone can be considered to be concentrated for moment calculations. The lever arm is the perpendicular distance from the moment reference axis to this centroid. This directly influence the moment capacity of the concrete section.

3.1 The Lever Arms Lax_{mrc} and Lay_{mrc}

Lever arms Lax_{mrc} and Lay_{mrc} for concrete are the distances from the x and y axis to the centroid of the compressive zone of concrete. This centroid, also known as the centre of gravity, is denoted by x̄ (x bar) and ȳ (y bar), and measured relative to their respective bases, x axis and y axis. This x̄ (x bar) and ȳ (y bar) are widely used in conventional mathematics and engineering to denote centre of gravity. This x̄ (x bar) and ȳ (y bar) distances are utilized to calculate the lever arms Lax_{mrc} and Lay_{mrc} for each x and y moment reference axes.

The centre of gravity varies depending on the shape of the compression zone area. Moreover, compressive area for each behaviour case of the column has unique and varied shapes. This means that the centroid of each shape must be determined using different methods and formulas for each behaviour case.

3.1.1 Behavior Case 1: $\alpha\delta b \leq b$ and $\alpha\delta h \leq h$

The following sections outline the approach for determining the lever arms Lax_{mrc} and Lay_{mrc} of the compression zone for the behaviour case 1, including a detailed calculation method and a worked example to illustrate each calculation step in detail.

3.1.1.1 Method of Calculation

The behaviour case 1 contains a compressive zone shaped as a triangle. The Equations 3.1 and 3.2 shows the basic equations to calculate the centroid of this triangular area. The Equations 3.3 and 3.4 shows the calculation formulas for the lever arms Lax_{mrc} and Lay_{mrc} for each x and y axis. This is illustrated in Figure 3.1.

$$\bar{\mathrm{x}} = \frac{1}{3} \cdot \alpha\delta b \tag{3.1}$$

$$\bar{\mathrm{y}} = \frac{1}{3} \cdot \alpha\delta h \tag{3.2}$$

$$Lax_{mrc} = (0.5 \cdot h) - \bar{\mathrm{y}} \tag{3.3}$$

$$Lay_{mrc} = (0.5 \cdot b) - \bar{\mathrm{x}} \tag{3.4}$$

Where,

$\bar{\mathrm{x}}$: The horizontal distance from the right vertical edge of the compression zone to its centroid.

$\bar{\mathrm{y}}$: The vertical distance from the top horizontal edge of the compression zone to its centroid.

$\alpha\delta b$ = Compression zone limit along the breadth.

$\alpha\delta h$ = Compression zone limit along the height.

Lax_{mrc}: Lever arm for concrete from moment reference axis x to the centroid of compression zone.

Lay_{mrc}: Lever arm for concrete from moment reference axis y to the centroid of compression zone.

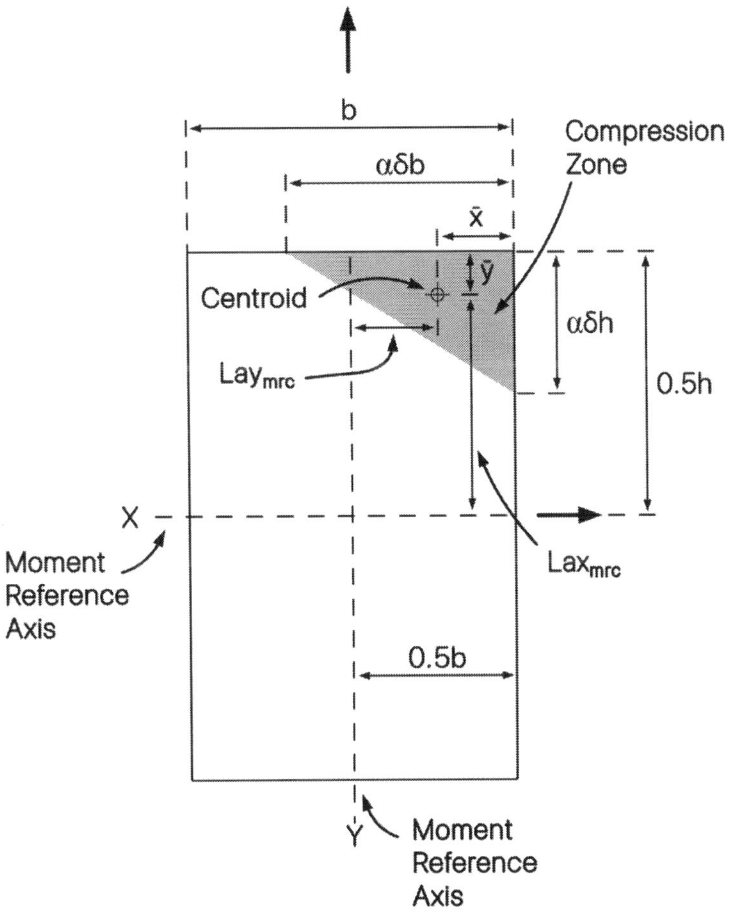

Figure 3.1: Lever arms and centroids for case 1

3.1.1.2 Example Calculation for Behaviour Case 1: $\alpha\delta b \leq b$, and $\alpha\delta h \leq h$

Given Data: h = 300 mm, b = 225 mm, $\alpha\delta h$ = 160 mm, $\alpha\delta b$ = 192 mm.

Method 1:

$$\bar{x} = \frac{1}{3} \cdot \alpha\delta b$$

$$= \frac{1}{3} \times 192 \text{ mm}$$

$$= 64 \text{ mm}$$

$$\bar{\text{y}} = \frac{1}{3} \cdot \alpha\delta h$$

$$= \frac{1}{3} \times 160 \text{ mm}$$

$$= 53.333 \text{ mm}$$

$$Lax_{mrc} = (0.5 \cdot h) - \bar{\text{y}}$$

$$= (0.5 \times 300 \text{ mm} - 53.333 \text{ mm}$$

$$= 96.667 \text{ mm}$$

$$Lay_{mrc} = (0.5 \cdot h) - \bar{\text{x}}$$

$$= (0.5 \times 225 \text{ mm} - 64 \text{ mm}$$

$$= 48.500 \text{ mm}$$

3.1.2 Behaviour Case 2: $\alpha\delta b > b$, and $\alpha\delta h \leq h$

The following sections outline the approach for determining the lever arms Lax_{mrc} and Lay_{mrc} of the compression zone for the behaviour case 2, including a detailed calculation method and a worked example to illustrate each calculation step in detail.

3.1.2.1 Method of Calculation

The behaviour case 2 contains a compressive zone shaped as a trapezium. The Equations 3.5 and 3.6 show the basic equations utilized to calculate the centroid of this trapezium. The Equations 3.7 and 3.8 show the formulas used to calculate the lever arms Lax_{mrc} and Lay_{mrc} for the x and y axes, respectively. This is illustrated in Figure 3.2.

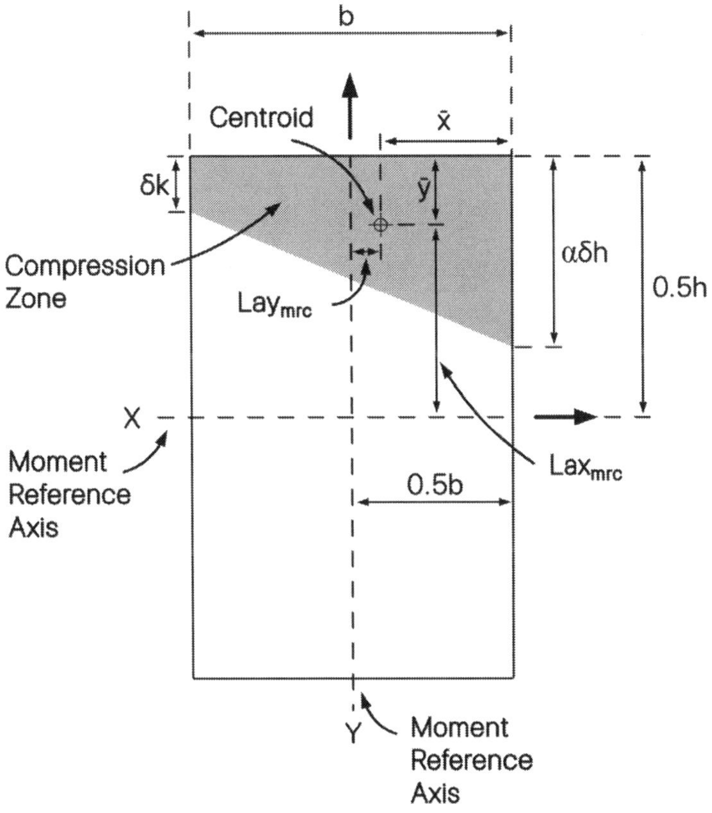

Figure 3.2: Lever arms and centroids for case 2

$$\bar{x} = \frac{b}{3} \cdot \frac{2\delta k + \alpha\delta h}{\delta k + \alpha\delta h} \tag{3.5}$$

$$\bar{y} = \frac{1}{3} \cdot \frac{\delta k^2 + (\alpha\delta h)^2 + (\alpha\delta h \cdot \delta k)}{\delta k + \alpha\delta h} \tag{3.6}$$

$$Lax_{mrc} = (0.5 \cdot h) - \bar{y} \tag{3.7}$$

$$Lay_{mrc} = (0.5 \cdot b) - \bar{x} \tag{3.8}$$

Where,

$\bar{x}$: The horizontal distance from the right vertical edge of the compression zone to its centroid.

$\bar{y}$: The vertical distance from the top horizontal edge of the compression

zone to its centroid.
$\alpha\delta h$ = Compression zone limit along the height.
δk: Length of the short side of compression area along the height.
Lax_{mrc}: Lever arm for concrete from moment reference axis x to the centroid of compression zone.
Lay_{mrc}: Lever arm for concrete from moment reference axis y to the centroid of compression zone.
b = Breadth of the column cross-section.
h = Height of the column cross-section.

3.1.2.2 Example Calculation for Behaviour case 2: $\alpha\delta b > b$, and $\alpha\delta h \leq h$

Given Data: $h = 300$ mm, $b = 225$ mm, $\alpha\delta h = 75$ mm, $\delta k = 25.368$ mm

Calculation:

$$\begin{aligned}
\bar{\mathrm{x}} &= \frac{b}{3} \cdot \frac{2\delta k + \alpha\delta h}{\delta k + \alpha\delta h} \\
&= \frac{225\,\text{mm}}{3} \times \frac{(2 \times (25.368\,\text{mm})) + 75\,\text{mm}}{25.368\,\text{mm} + 75\,\text{mm}} \\
&= 93.956\,\text{mm} \\
\bar{\mathrm{y}} &= \frac{1}{3} \cdot \frac{\delta k^2 + (\alpha\delta h)^2 + (\alpha\delta h \cdot \delta k)}{\delta k + \alpha\delta h} \\
&= \frac{1}{3} \times \frac{(25.368\,\text{mm})^2 + (75\,\text{mm})^2 + (75\,\text{mm} \times 25.368\,\text{mm})}{25.368\,\text{mm} + 75\,\text{mm}} \\
&= 27.137\,\text{mm} \\
Lax_{mrc} &= (0.5 \cdot h) - \bar{\mathrm{y}} \\
&= (0.5 \times 300\,\text{mm}) - 27.137\,\text{mm} \\
&= 122.863\,\text{mm}
\end{aligned}$$

$$Lay_{mrc} = (0.5 \cdot b) - \bar{x}$$

$$= (0.5 \times 225\,\text{mm}) - 93.956\,\text{mm}$$

$$= 18.544\,\text{mm}$$

3.1.3 Behaviour Case 3: $\alpha\delta b \leq b$, and $\alpha\delta h > h$

The following sections outline the approach for determining the lever arms Lax_{mrc} and Lay_{mrc} of the compression zone for the behaviour case 3, including a detailed calculation method and a worked example to illustrate each calculation step in detail.

3.1.3.1 Method of Calculation

The behaviour case 3 contains a compressive zone shaped as a trapezium. The Equations 3.9 and 3.10 show the basic expressions utilized to calculate the centroid of this trapezium. The Equations 3.11 and 3.12 show the formulas utilized to calculate the lever arms Lax_{mrc} and Lay_{mrc} for the x and y axes, respectively. This is illustrated in Figure 3.3.

$$\bar{x} = \frac{1}{3} \cdot \frac{\delta v^2 + (\alpha\delta b)^2 + (\alpha\delta b \cdot \delta v)}{\delta v + \alpha\delta b} \tag{3.9}$$

$$\bar{y} = \frac{b}{3} \cdot \frac{2\delta v + \alpha\delta b}{\delta v + \alpha\delta b} \tag{3.10}$$

$$Lax_{mrc} = (0.5 \cdot h) - \bar{y} \tag{3.11}$$

$$Lay_{mrc} = (0.5 \cdot b) - \bar{x} \tag{3.12}$$

Where,

$\bar{x}$: The horizontal distance from the right vertical edge of the compression zone to its centroid.

$\bar{y}$: The vertical distance from the top horizontal edge of the compression

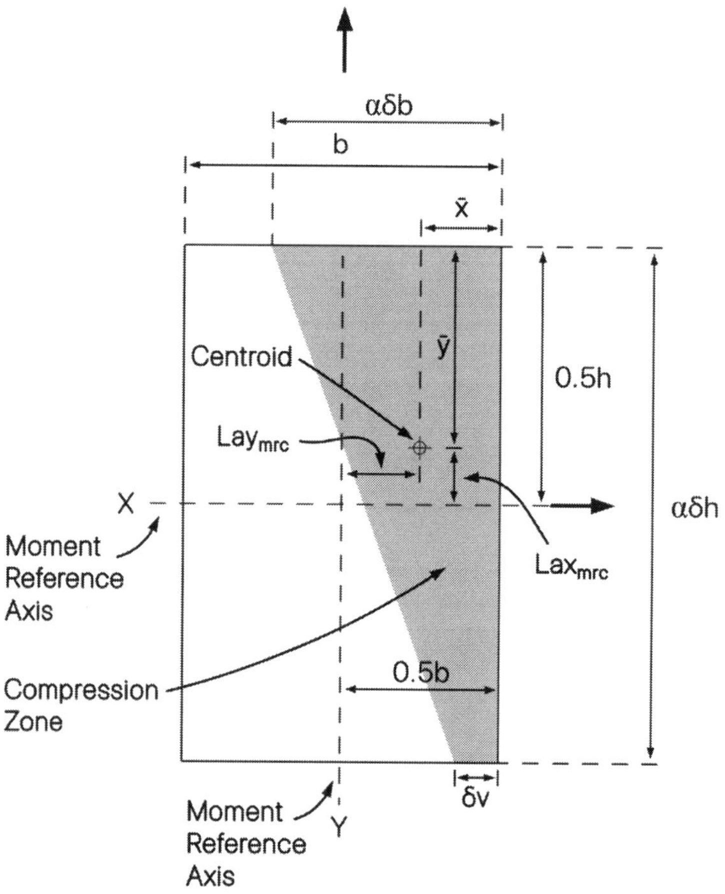

Figure 3.3: Lever arms and centroids for case 3

zone to its centroid.

$\alpha\delta h$ = Compression zone limit along the height.

$\alpha\delta b$ = Compression zone limit along the breadth.

δv: Length of the short side of compression area along the breadth.

Lax_{mrc}: Lever arm for concrete from moment reference axis x to the centroid of compression zone.

Lay_{mrc}: Lever arm for concrete from moment reference axis y to the centroid of compression zone.

b = Breadth of the column cross-section.

h = Height of the column cross-section.

3.1.3.2 Example Calculation for Behaviour Case 3: $\alpha\delta b \leq b$, and $\alpha\delta h > h$

Given Data: $h = 300$ mm, $b = 225$ mm, $\alpha\delta b = 170$ mm, $\delta v = 34$ mm

Calculation:

$$\bar{x} = \frac{1}{3} \cdot \frac{\delta v^2 + (\alpha\delta b)^2 + (\alpha\delta b \cdot \delta v)}{\delta v + \alpha\delta b}$$

$$= \frac{1}{3} \times \frac{(34\,\text{mm})^2 + (170\,\text{mm})^2 + (170\,\text{mm} \times 34\,\text{mm})}{34\,\text{mm} + 170\,\text{mm}}$$

$$= 58.556\,\text{mm}$$

$$\bar{y} = \frac{b}{3} \cdot \frac{2\delta_v + \alpha\delta b}{\delta v + \alpha\delta_b}$$

$$= \frac{225\,\text{mm}}{3} \times \frac{2(34\,\text{mm}) + 170\,\text{mm}}{34\,\text{mm} + 170\,\text{mm}}$$

$$= 87.500\,\text{mm}$$

$$Lax_{mrc} = (0.5 \cdot h) - \bar{y}$$

$$= (0.5 \times 300\,\text{mm}) - 87.500\,\text{mm}$$

$$= 62.500\,\text{mm}$$

$$Lay_{mrc} = (0.5 \cdot b) - \bar{x}$$

$$= (0.5 \times 225\,\text{mm}) - 58.556\,\text{mm}$$

$$= 53.944\,\text{mm}$$

3.1.4 Behaviour Case 4: $\alpha\delta b > b$, and $\alpha\delta h > h$

The following sections outline the approach for determining the lever arms Lax_{mrc} and Lay_{mrc} of the pentagonal compression zone for the behaviour case 4, including a detailed calculation method and a worked example to illustrate each calculation step in detail.

3.1.4.1 Method of Calculation

Before calculating the lever arm, the centroid of the pentagon must be determined. To do this, the basic principle behind it must be understood. The centroid of a physical object is the weighted average of the moments of all contributing areas. Simply, if the object is sliced into fine strips, the sum of the weight of each strip multiplied by its centroid, divided by the total area, gives the centroid of the object. This is shown in the Equations 3.13 and 3.14.

$$\bar{\mathrm{x}} = \frac{\sum(A_i \cdot x_i)}{\sum A_i} \tag{3.13}$$

$$\bar{\mathrm{y}} = \frac{\sum(A_i \cdot y_i)}{\sum A_i} \tag{3.14}$$

Where,
$\bar{\mathrm{x}}$: Centroid of the whole object along x axis.
$\bar{\mathrm{y}}$: Centroid of the whole object along y axis.
A_i: Area of each slice.
x_i: Centroid of each slice along x axis from right surface.
y_i: Centroid of each slice along y axis from top surface.

In the same manner, this can be applied to the rectangular column. Divide the whole rectangular cross-section into two slices of a pentagonal shape and a triangle. Then, remove the triangular non-compressive area from the cross section to calculate the centroid of the remaining pentagonal shape. This is shown in Equations 3.18 and 3.19. The total cross-sectional area of the column, the non-compressive triangular area, and the pentagonal compressive zone are shown in Equations 3.15 to 3.17.
Moreover, the distance to centroid of the pentagonal compression area is determined by subtracting the centroids of the triangular area from the rectangular cross-section of the column. This distance to centroid

of the pentagonal compression zone from the top and right surfaces of the column along the x axis and y axis (x_{Pent} and y_{Pent}) is shown in Equations 3.20 and 3.21. The Lever arms Lax_{mrc} and Lay_{mrc} are shown in Equations 3.22 and 3.23. This is further illustrated in Figure 3.4.

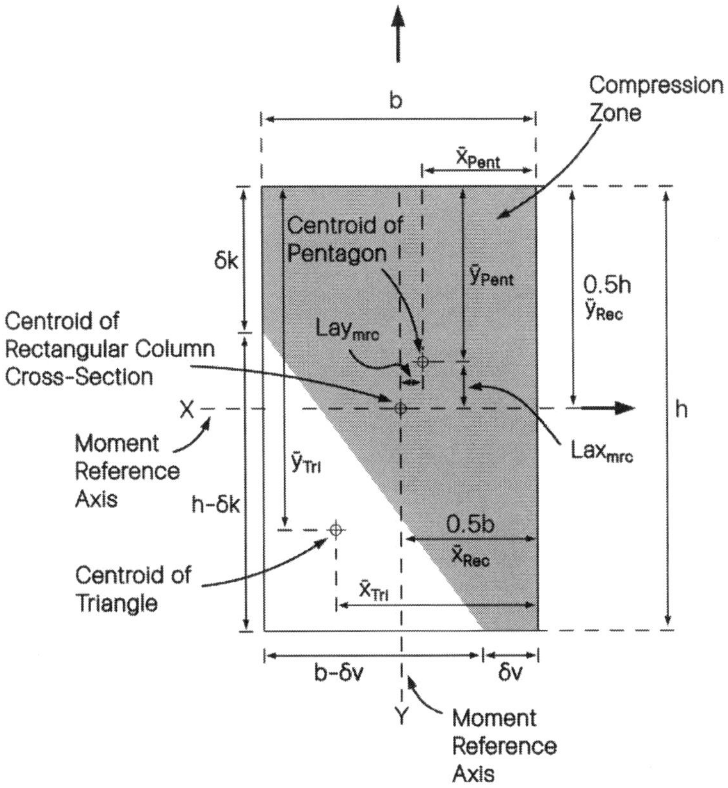

Figure 3.4: Lever arms and centroids for case 4

$$A_{Tri} = 0.5 \cdot (h - \delta k) \cdot (b - \delta v) \tag{3.15}$$

$$A_{Total} = h \cdot b \tag{3.16}$$

$$A_{Pent} = A_{Total} - A_{Tri} \tag{3.17}$$

$$\bar{x}_{Rec} = \frac{(\bar{x}_{Pent} \cdot A_{Pent}) + (\bar{x}_{Tri} \cdot A_{Tri})}{A_{Total}} \tag{3.18}$$

$$\bar{y}_{Rec} = \frac{(\bar{y}_{Pent} \cdot A_{Pent}) + (\bar{y}_{Tri} \cdot A_{Tri})}{A_{Total}} \tag{3.19}$$

$$\bar{x}_{Pent} = \frac{(\bar{x}_{Rec} \cdot A_{Total}) - (\bar{x}_{Tri} \cdot A_{Tri})}{A_{Pent}} \tag{3.20}$$

$$\bar{y}_{Pent} = \frac{(\bar{y}_{Rec} \cdot A_{Total}) - (\bar{y}_{Tri} \cdot A_{Tri})}{A_{Pent}} \tag{3.21}$$

$$Lax_{mrc} = 0.5h - \bar{y}_{Pent} \tag{3.22}$$

$$Lay_{mrc} = 0.5b - \bar{x}_{Pent} \tag{3.23}$$

Where,

$\bar{x}_{Rec}$: Distance to centroid of the rectangular column cross-section from the right surface of the column along the x axis.
$\bar{y}_{Rec}$: Distance to centroid of the rectangular column cross-section from the top surface of the column along the y axis.
$\bar{x}_{Pent}$: Distance to centroid of the pentagonal compression zone from the right surface of the column along the x axis.
$\bar{y}_{Pent}$: Distance to centroid of the pentagonal compression zone from the top surface of the column along the y axis.
$\bar{x}_{Tri}$: Distance to centroid of the triangular non-compressive zone from the right surface of the column along the x axis.
$\bar{y}_{Tri}$: Distance to centroid of the triangular non-compressive zone from the top surface of the column along the y axis.
A_{Pent}: Area of the pentagonal compression zone.
A_{Tri}: Area of the triangular non-compressive zone.
A_{Total}: Total area of the rectangular cross-section.
Lax_{mrc}: Lever arm for concrete from moment reference axis x to the centroid of compression zone.
Lay_{mrc}: Lever arm for concrete from moment reference axis y to the centroid of compression zone.

Moreover, detailed equations were utilized to calculate $\bar{x}_{Rec}$ and $\bar{y}_{Rec}$, as shown in Equations 3.18 and 3.19, which include all necessary components to derive $\bar{x}_{Pent}$ and $\bar{y}_{Pent}$. However, since the column has a rectangular shape, a simplified version can also be utilized for the calculation, as shown in Equations 3.24 and 3.25.

$$\bar{x}_{Rec} = 0.5 \cdot b \tag{3.24}$$

$$\bar{y}_{Rec} = 0.5 \cdot h \tag{3.25}$$

The centroid of the non-compressive right-angled triangle is shown in Equations 3.26 and 3.27.

$$\bar{x}_{Tri} = b - \left(\frac{b - \delta v}{3}\right) \tag{3.26}$$

$$\bar{y}_{Tri} = h - \left(\frac{h - \delta k}{3}\right) \tag{3.27}$$

Where,
$\bar{x}_{Tri}$: Distance to centroid of the triangular non-compressive zone from the right surface of the column along the x axis.
$\bar{y}_{Tri}$: Distance to centroid of the triangular non-compressive zone from the top surface of the column along the y axis.
b = Breadth of the column cross-section.
h = Height of the column cross-section.
δk = Length of the long side of compression area along the height.
δv = Length of the short side of compression area along the breadth.

3.1.4.2 Example Calculation for Behaviour Case 4: $\alpha\delta b > b$, and $\alpha\delta h > h$

Given Data: $h = 300$ mm, $b = 225$ mm, $\delta v = 68$ mm, $\delta k = 126.836$ mm, $A_{Pent} = 53,906.783$ mm

Calculation:

$$\begin{aligned} A_{Total} &= h \cdot b \\ &= 300\,\text{mm} \times 225\,\text{mm} \end{aligned}$$

$$= 67500.000\,\text{mm}^2$$

$$A_{Tri} = A_{Total} - A_{Pent}$$

$$= 67500\,\text{mm}^2 - 53906.783\,\text{mm}^2$$

$$= 13593.217\,\text{mm}^2$$

$$\bar{\text{x}}_{Tri} = b - \frac{b - \delta v}{3}$$

$$= 225\,\text{mm} - \frac{225\,\text{mm} - 68\,\text{mm}}{3}$$

$$= 172.667\,\text{mm}$$

$$\bar{\text{y}}_{Tri} = h - \frac{h - \delta k}{3}$$

$$= 300\,\text{mm} - \frac{300\,\text{mm} - 126.836\,\text{mm}}{3}$$

$$= 242.279\,\text{mm}$$

$$\bar{\text{x}}_{Rec} = \frac{b}{2} = \frac{225\,\text{mm}}{2} = 112.500\,\text{mm}$$

$$\bar{\text{y}}_{Rec} = \frac{h}{2} = \frac{300\,\text{mm}}{2} = 150.000\,\text{mm}$$

$$\bar{\text{x}}_{Pent} = \frac{(\bar{\text{x}}_{Rec} \cdot A_{Total}) - (\bar{\text{x}}_{Tri} \cdot A_{Tri})}{A_{Pent}}$$

$$= \frac{(112.5\,\text{mm} \times 67500\,\text{mm}^2) - (172.667\,\text{mm} \times 13593.217\,\text{mm}^2)}{53906.783\,\text{mm}^2}$$

$$= 97.328\,\text{mm}$$

$$\bar{\text{y}}_{Pent} = \frac{(\bar{\text{y}}_{Rec} \cdot A_{Total}) - (\bar{\text{y}}_{Tri} \cdot A_{Tri})}{A_{Pent}}$$

$$= \frac{(150.000\,\text{mm} \cdot 67500.000\,\text{mm}^2) - (242.279\,\text{mm} \cdot 13593.217\,\text{mm}^2)}{53906.783\,\text{mm}^2}$$

$$= 126.731\,\text{mm}$$

$$Lax_{mrc} = 0.5h - \bar{\text{y}}_{Pent}$$

$$= 0.5 \times 300\,\text{mm} - 126.731\,\text{mm}$$
$$= 23.269\,\text{mm}$$

$$Lay_{mrc} = 0.5b - \bar{\text{x}}_{Pent}$$
$$= 0.5 \times 225\,\text{mm} - 97.329\,\text{mm}$$
$$= 15.171\,\text{mm}$$

3.2 The Lever Arms Lax_{mrci} and Lay_{mrci}

Distances from the centroids of the concrete compression zone to the inclined (rotated) moment reference axes (x and y) are determined by accounting for the inclination angle (θ) of these axes relative to the non-inclined (unrotated) axes. The rotated axes intersect at the centroid of the section.

Two calculation methods are considered.

Method 1: Uses direct projection of horizontal and vertical distances onto the inclined axes.

Method 2: Uses geometric relationships to calculate the distances, taking the axis inclination into account.

3.2.1 Method 1

In this method, the distances from the centroid of the compression zone to the inclined moment reference axes (rotated x and y axes) by directly projecting the horizontal $\bar{\text{x}}$ (x bar) and vertical $\bar{\text{y}}$ (y bar) distances onto the inclined axes. The projections are calculated by aligning the rotated axes with the neutral axis inclination (θ). Consequently, utilizing trigonometric relationships, taking into account the angle of inclination (θ) of the axes relative to the unrotated x axis. This is shown in Equations 3.28 and 3.29, and further illustrated in Figure 3.5.

3.2.1.1 Method of Calculation

$$Lax_{mrci} = (Lax_{mrc} \cdot \cos(\theta)) + (Lay_{mrc} \cdot \sin(\theta)) \tag{3.28}$$

$$Lay_{mrci} = (Lay_{mrc} \cdot \cos(\theta)) - (Lax_{mrc} \cdot \sin(\theta)) \tag{3.29}$$

Where,

Lax_{mrci}: Distance from the centroid of the compression zone to inclined (rotated) moment reference axis x.

Lay_{mrci}: Distance from the centroid of the compression zone to inclined (rotated) moment reference axis y.

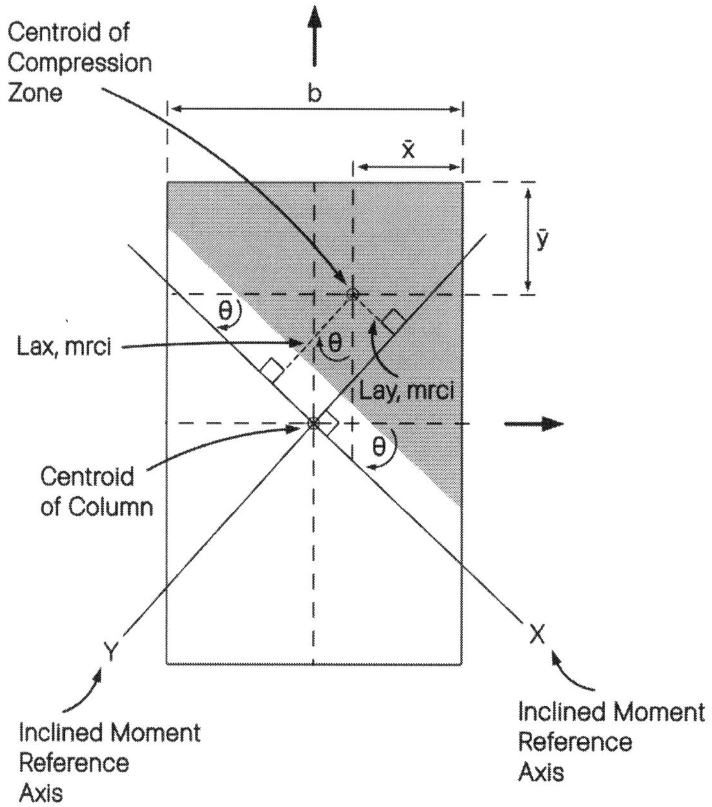

Figure 3.5: Lever arms Lax_{mrci} *and* Lay_{mrci} for method 1

3.2.1.2 Example Calculation

Given Data: $\theta = 39.805571092°$, $Lax_{mrc} = 96.667$ mm, $Lay_{mrc} = 48.500$ mm.

Calculation:

$$\begin{aligned}
Lax_{mrci} &= (Lax_{mrc} \cdot \cos(\theta)) + (Lay_{mrc} \cdot \sin(\theta)) \\
Lay_{mrci} &= (Lay_{mrc} \cdot \cos(\theta)) - (Lax_{mrc} \cdot \sin(\theta)) \\
\cos(\theta) &= \cos(39.805571092°) = 0.76822128 \\
\sin(\theta) &= \sin(39.805571092°) = 0.6401844 \\
Lax_{mrci} &= (96.667\,\text{mm} \times 0.76822128) + (48.500\,\text{mm} \times 0.6401844) \\
&= 105.310\,\text{mm} \\
Lay_{mrci} &= (48.500\,\text{mm} \times 0.76822128) - (96.667\,\text{mm} \times 0.6401844) \\
&= -24.630\,\text{mm}
\end{aligned}$$

Note: When using spreadsheet software, ensure that the output of the COS or SIN function is correctly interpreted. Use $COS(RADIANS(\theta))$ or $SIN(RADIANS(\theta))$ to convert degrees to radians, as some software expects angles in radians by default.

3.2.2 Method 2

This method employs geometric principles to calculate distances directly from concrete compression zone centroids to the inclined (rotated) axes. By aligning the rotated axes with the neutral axis inclination (θ), distances are calculated through sequential steps involving projections and trigonometric relationships.

The distances from each rotated axis are obtained using basic geometric and trigonometric principles, as follows.

3.2.2.1 Distance from Inclined (Rotated) x Axis (Lax_{mrci})

This section calculates the horizontal distances Lax_{mrci} from compression zone centroids to the inclined (rotated) x-axis. Using basic geometry, the distances are derived by projecting compression zone centroid coordinates onto the inclined axis, considering the inclination angle (θ) and the geometric relationships within the section.

3.2.2.1.1 Method of Calculation

The following shows the basic method to calculate the distance from the reinforcement centroid to the inclined (rotated) x-axis. Moreover, Equations 3.30 to 3.34 show the final derived expressions for each calculation step. This is illustrated in Figure 3.6.

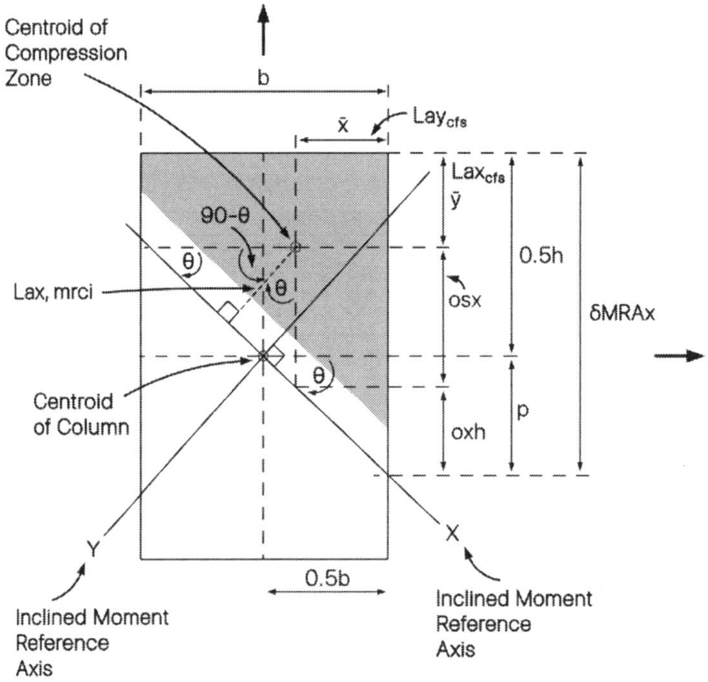

Figure 3.6: Lever arm Lax_{mrci} for method 2

Derivation of Lax_{mrci}

$$\cos(\theta) = \frac{\text{adjacent}}{\text{hypotenuse}}$$

Therefore,

$$\cos(\theta) = \frac{Lax_{mrci}}{osx}$$

$$Lax_{mrci} = \cos(\theta) \cdot osx$$

Derivation of osx

$$osx = \delta MRAx - (oxh + \bar{\text{y}})$$

Derivation of oxh

$$\tan(\theta) = \frac{\text{opposite}}{\text{adjacent}}$$

$$\tan(\theta) = \frac{oxh}{\bar{\text{x}}}$$

$$oxh = \tan(\theta) \cdot \bar{\text{x}}$$

Derivation of $\delta MRAx$

$$\delta MRAx = (0.5 \cdot h) + P$$

Derivation of P

$$\tan(\theta) = \frac{\text{opposite}}{\text{adjacent}}$$

$$\tan(\theta) = \frac{P}{0.5 \cdot b}$$

$$P = \tan(\theta) \cdot (0.5 \cdot b)$$

Summary of derived equations

$$Lax_{mrci} = \cos(\theta) \cdot osx \tag{3.30}$$

$$osx = \delta MRAx - (oxh + \bar{y}) \tag{3.31}$$

$$oxh = \tan(\theta) \cdot \bar{x} \tag{3.32}$$

$$\delta MRAx = (0.5 \cdot h) + P \tag{3.33}$$

$$P = \tan(\theta) \cdot (0.5 \cdot b) \tag{3.34}$$

Where,

Lax_{mrci}: Distance from the centroid of the compression zone to inclined (rotated) moment reference axis x
$\bar{x}$: Centroid of compression zone along breadth (b)
$\bar{y}$: Centroid of the compression zone along height (h)
h: Height of column
b: Breadth of column

3.2.2.1.2 Example Calculation

Given Data: $h = 300$ mm, $b = 225$ mm, $\bar{x} = 64$ mm, $\bar{y} = 53.333$ mm, $\theta = 39.805571092°$

Calculation:

Calculation of P

$$\begin{aligned} P &= \tan(\theta) \cdot 0.5 \cdot b \\ &= \tan(39.805571092°) \times (0.5 \times 225\,\text{mm}) \\ &= 0.833333333 \times 112.5\,\text{mm} \\ &= 93.75\,\text{mm} \end{aligned}$$

Calculation of $\delta MRAx$

$$\begin{aligned}\delta MRAx &= (0.5 \cdot h) + P \\ &= (0.5 \times 300)\,\text{mm} + 93.75\,\text{mm} \\ &= 150\,\text{mm} + 93.75\,\text{mm} \\ &= 243.75\,\text{mm}\end{aligned}$$

Calculation of oxh

$$\begin{aligned}oxh &= \tan(\theta) \cdot \bar{\text{x}} \\ &= \tan(39.805571092^\circ) \times 64.000\,\text{mm} \\ &= 0.833333333 \times 64.000\,\text{mm} \\ &= 53.334\,\text{mm}\end{aligned}$$

Calculation of osx

$$\begin{aligned}osx &= \delta MRAx - (oxh + \bar{\text{y}}) \\ &= 243.75\,\text{mm} - (53.334\,\text{mm} + 53.333\,\text{mm}) \\ &= 243.75\,\text{mm} - 106.667\,\text{mm} \\ &= 137.083\,\text{mm}\end{aligned}$$

Calculation of Lax_{mrci}

$$\begin{aligned}Lax_{mrci} &= \cos(\theta) \cdot osx \\ &= \cos(39.805571092^\circ) \times 137.083\,\text{mm} \\ &= 0.76822128 \times 137.083\,\text{mm} \\ &= 105.310\,\text{mm}\end{aligned}$$

3.2.2.2 Distance from Inclined (Rotated) y Axis (Lay_{mrci})

This section calculates the horizontal distances Lay_{mrci} from compression zone centroids to the inclined (rotated) y-axis. Using basic geometry, the distances are derived by projecting compression zone centroid coordinates onto the inclined axis, considering the inclination angle (θ) and the geometric relationships within the section.

3.2.2.2.1 Method of Calculation

The following shows the basic method to calculate the distance from the compression zone centroid to the inclined (rotated) x-axis. Moreover, Equations 3.35 to 3.39 show the final derived expressions for each calculation step. This is further illustrated in Figures 3.7 and 3.8. The inclined moment reference axis (rotated x axis) is aligned to the neutral axis, sharing the same inclination angle (θ).

The following figures and centroid locations may or may not be to scale and are for illustrative purposes only.

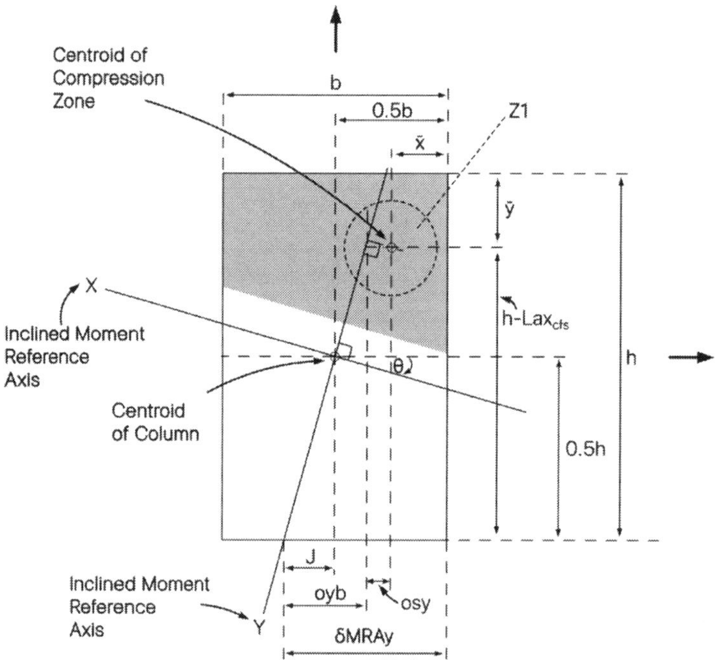

Figure 3.7: Lever arm Lay_{mrci} for method 2

Derivation of Lay_{mrci}

$$\cos(\theta) = \frac{\text{adjacent}}{\text{hypotenuse}}$$

$$\cos(\theta) = \frac{Lay_{mrci}}{osy}$$

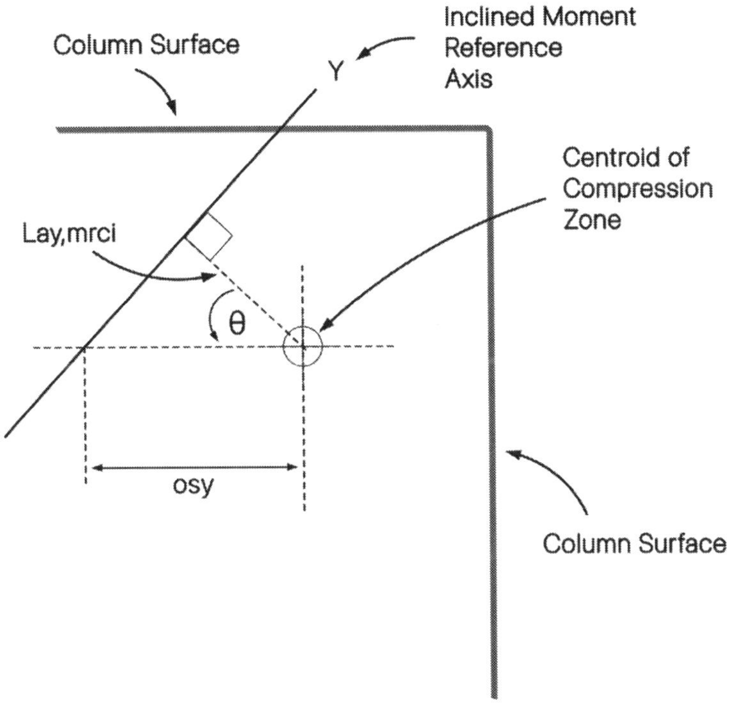

Figure 3.8: Magnification Z1

$$Lay_{mrci} = \cos(\theta) \cdot osy$$

Derivation of osy

$$osy = \delta MRAy - (oyb + \bar{x})$$

Derivation of oyb

$$\tan(\theta) = \frac{\text{opposite}}{\text{adjacent}}$$

$$\tan(\theta) = \frac{oyb}{h - \bar{y}}$$

$$oyb = \tan(\theta) \cdot (h - \bar{y})$$

Derivation of $\delta MRAy$

$$\delta MRAy = (0.5 \cdot b) + J$$

Derivation of J

$$\tan(\theta) = \frac{\text{opposite}}{\text{adjacent}}$$

$$\tan(\theta) = \frac{J}{0.5 \cdot h}$$

$$J = \tan(\theta) \cdot (0.5 \cdot h)$$

Summary of derived equations

$$Lay_{mrci} = \cos(\theta) \cdot osy \tag{3.35}$$

$$osy = \delta MRAy - (oyb + \bar{x}) \tag{3.36}$$

$$oyb = \tan(\theta) \cdot (h - \bar{y}) \tag{3.37}$$

$$\delta MRAy = (0.5 \cdot b) + J \tag{3.38}$$

$$J = \tan(\theta) \cdot (0.5 \cdot h) \tag{3.39}$$

Where,

Lay_{mrci}: Distance from the centroid of the compression zone to inclined (rotated) moment reference axis y.

$\bar{x}$: Centroid of the compression zone along breadth (b).

$\bar{y}$: Centroid of the compression zone along height (h).

h: Height of column.

b: Breadth of column.

θ: Inclination angle (rotation angle) of the inclined (rotated) axis relative to the non-inclined (unrotated) axis.

3.2.2.2.2 Example Calculation

Given Data: $h = 300\,\text{mm}$, $b = 225\,\text{mm}$, $\bar{x} = 64\,\text{mm}$, $\bar{y} = 53.333\,\text{mm}$, $\theta = 39.805571092°$

Calculation:

Calculation of J

$$\begin{aligned} J &= \tan(\theta) \cdot 0.5 \cdot h \\ &= \tan(39.805571092°) \times (0.5 \times 300\,\text{mm}) \\ &= 0.833333333 \times 150\,\text{mm} \\ &= 125\,\text{mm} \end{aligned}$$

Calculation of $\delta MRAy$

$$\begin{aligned} \delta MRAy &= 0.5 \cdot b + J \\ &= (0.5 \times 225\,\text{mm}) + 125\,\text{mm} \\ &= 112.5\,\text{mm} + 125\,\text{mm} \\ &= 237.5\,\text{mm} \end{aligned}$$

Calculation of oyb

$$\begin{aligned} oyb &= \tan(\theta) \cdot (h - \bar{\text{y}}) \\ &= \tan(39.805571092°) \times (300\,\text{mm} - 53.333\,\text{mm}) \\ &= 0.833333333 \times 246.667\,\text{mm} \\ &= 205.556\,\text{mm} \end{aligned}$$

Calculation of osy

$$\begin{aligned} osy &= \delta MRAy - (oyb + \bar{\text{x}}) \\ &= 237.5\,\text{mm} - (205.556\,\text{mm} + 64\,\text{mm}) \\ &= 237.5\,\text{mm} - 269.556\,\text{mm} \\ &= -32.056\,\text{mm} \end{aligned}$$

Calculation of Lay_{mrci}

$$\begin{aligned} Lay_{mrci} &= \cos(\theta) \cdot osy \\ &= \cos(39.805571092°) \times (-32.056\,\text{mm}) \\ &= 0.76822128 \times (-32.056\,\text{mm}) \end{aligned}$$

$$= -24.630\,\text{mm}$$

3.2.3 Summary of Calculated Distances

The following Table 3.1 shows the summary of the distances (Lax_{mrci} and Lay_{mrci}) from the centroid of concrete compression zone to the inclined moment reference axes x and y.

Note that the results obtained from Method 2 are consistent with those from Method 1, as both methods calculate the same parameters.

Table 3.1: Summary of the distances

Distance Type	**Distance**
Lax_{mrci}	+105.31 mm
Lay_{mrci}	-24.63 mm

Chapter 4:

Resistive Axial Force and Moments in Concrete

4.1 Strain (ε) and Stress (σ) for the Concrete Compression Zone

The simplified compression zone, uses 0.8λ as mentioned in Eurocode 2 (EN 1992-1-1), clause 3.1.7(3). This approach assumes that the stress is uniform throughout the compression block. It also assumes zero strain at the neutral axis, and no compressive stress outside the simplified compression zone.

These assumptions are only valid for design purposes of concrete compression zone. Moreover, these are engineering approximations, not strict physical truths.

As a result, it is not required to calculate strain at the centroid of the compression zone or determine stress based on strain. Instead, stress is calculated directly by applying partial safety factors and other adjustment factors to the stress of the concrete.

It is worth noting that in a later section of this book, when calculating reinforcement stress, the actual strain is first determined, and consequently, the stress is derived from it, rather than calculating the stress directly with simplified methods.

4.1.1 Method of Calculation

The Equation 4.1 shows the design strength (fcd) of concrete which is utilized as design stress (σcd) of the concrete compression zone. This is stated in the Clause 3.1.6 (1) in Eurocode 2 (EN 1992-1-1). The Gamma (γ) and Alpha (α) are Greek letters utilized for notation.

$$f_{cd} = \frac{\alpha_{cc} \cdot f_{ck}}{\gamma_c} \tag{4.1}$$

Where,
f_{ck} = Characteristic compressive strength of concrete.
f_{cd} = Design compressive strength of concrete.
γ_c = Partial safety factor for concrete (Typically, $\gamma_c = 1.5$).
α_{cc} = 0.85 (This is the coefficient accounting for long-term effects on compressive strength and unfavourable load application effects. The standard recommended value in Eurocode is 1.0 in shear situations. Nevertheless, this depends on the national annex of your country. For a conservative approach, and since columns experience both flexure and shear, you may use 0.85 as α_{cc}, as suggested by the UK National Annex.)

4.1.2 Example Calculation

Given Data: $f_{ck} = 30\,\text{N/mm}^2$, $\alpha_{cc} = 0.85$, $\gamma_c = 1.5$.

Calculation:

$$\begin{aligned} f_{cd} &= \frac{\alpha_{cc} \cdot f_{ck}}{\gamma_c} \\ &= \frac{0.85 \cdot 30\,\text{N/mm}^2}{1.5} \\ &= 17\,\text{N/mm}^2 \end{aligned}$$

4.2 Resistive axial compressive force in concrete (N_c)

The resistive axial compressive force in concrete (N_c), also known as the axial compressive force in concrete, is the force developed within the concrete when it resists compressive stresses. This force arises from the ability of the concrete to withstand compressive loads, effectively balancing the internal forces within the concrete section.

4.2.1 Method of Calculation

In some practices, a common assumption is that the entire compressive zone of concrete contributes to resistance. However, that approach does not account for the displacement of concrete by reinforcement in the compressive zone. As a result, the calculation effectively overestimates the compressive resistance by treating the reinforcement-occupied regions as if concrete were present. This introduces a degree of inaccuracy.

To address this issue, the reinforcement area can be deducted from the concrete, which reduces the area of the compression zone. However, this slightly shifts the centroid of the compression zone, introducing a minor inaccuracy. To mitigate this, the area of the reinforcement is not deducted at this stage. To compensate for that, the stress contribution of the concrete area at the reinforcement location is reduced during the calculation of reinforcement forces in a subsequent section. This method resolves the inaccuracies. Thereby, the following axial force of concrete (N_c) calculation does not reduce the reinforcement area. The compressive force in concrete (N_c) is shown in Equation 4.2.

The Gamma (γ), Eta (η), and Omega (ω), are Greek letters utilized for notation.

$$N_c = \omega \cdot \eta \cdot f_{cd} \cdot A_{cc} \tag{4.2}$$

Where,
η = Partial safety factor applied to the material strength (Typically 1.0)
A_{cc} = Area of compressive concrete.
ω = 0.9 (Reduction factor to account for the non-linear distribution when compression zone decreases toward the extreme compression fibre (EC2, Clause 3.1.7 (3))).
fcd = Design compressive strength of concrete.

4.2.2 Example Calculation

Behaviour Case 1: $\alpha\delta b \leq b$, and $\alpha\delta h \leq h$

Given Data: $A_{cc} = 15360\,\text{mm}^2$, $\omega = 0.9$, $\eta = 1.0$, $f_{cd} = 17\,\text{N/mm}^2$.

Calculation:

$$\begin{aligned} N_c &= \omega \cdot \eta \cdot f_{cd} \cdot A_{cc} \\ &= 0.9 \times 1 \times 17\,\text{N/mm}^2 \times 15360\,\text{mm}^2 \\ &= 235,008\,\text{N} = 235.008\,\text{kN} \end{aligned}$$

Behaviour Case 2: $\alpha\delta b > b$, and $\alpha\delta h \leq h$

Given Data: $A_{cc} = 11,291.360\,\text{mm}^2$, $\omega = 0.9$, $\eta = 1.0$, $f_{cd} = 17\,\text{N/mm}^2$.

Calculation:

$$\begin{aligned} N_c &= \omega \cdot \eta \cdot f_{cd} \cdot A_{cc} \\ &= 0.9 \times 1 \times 17\,\text{N/mm}^2 \times 11,291.360\,\text{mm}^2 \\ &= 172,757.808\,\text{N} = 172.758\,\text{kN} \end{aligned}$$

Behaviour Case 3: $\alpha\delta b \leq b$, and $\alpha\delta h > h$

Given Data: $A_{cc} = 30,600\,\text{mm}^2$, $\omega = 0.9$, $\eta = 1.0$, $f_{cd} = 17\,\text{N/mm}^2$.

Calculation:

$$\begin{aligned} N_c &= \omega \cdot \eta \cdot f_{cd} \cdot A_{cc} \\ &= 0.9 \times 1 \times 17\,\text{N/mm}^2 \times 30,600\,\text{mm}^2 \\ &= 468,180\,\text{N} = 468.180\,\text{kN} \end{aligned}$$

Behaviour Case 4: $\alpha\delta b > b$, **and** $\alpha\delta h > h$

Given Data: $A_{cc} = 53,906.783\,\text{mm}^2$, $\omega = 0.9$, $\eta = 1.0$, $f_{cd} = 17\,\text{N/mm}^2$.

Calculation:

$$\begin{aligned} N_c &= \omega \cdot \eta \cdot f_{cd} \cdot A_{cc} \\ &= 0.9 \times 1 \times 17\,\text{N/mm}^2 \times 53,906.783\,\text{mm}^2 \\ &= 824,773.778\,\text{N} = 824.774\,\text{kN} \end{aligned}$$

4.3 Resistive Bending Moments in Concrete

The resistive bending moments in concrete, denoted as M_{cxn} and M_{cyn} relative to non-inclined moment reference axis (unrotated axes), and M_{cxi} and M_{cyi} relative to inclined moment reference axis (rotated axes), represent the contributions of the concrete section in resisting bending moments about the x and y axis, respectively. These moments are critical in evaluating the overall moment capacity of the column.

In a reinforced concrete section under biaxial bending, the concrete in the compressive zone plays a key role in resisting bending moments. Unlike steel, which primarily resists both tension and compression through reinforcement bars, concrete contributes to the bending resistance through its ability to bear compressive stresses across a compressive area. The tensile resistance of the concrete is insignificant. Thereby, it is not considered for moment resistance calculations.

By calculating resistive bending moments in concrete for each axis, reinforced concrete members are designed to resist the combined effects of biaxial bending moments while ensuring that both the concrete and steel reinforcement are utilized effectively within their respective capacities. This ensures stability, safety, and efficiency in structural performance.

There are two primary approaches to calculate these resistive bending moments, distinguished by the orientation of the moment reference axes relative to the neutral axis. In the first method, referred to as the non-inclined or unrotated axes approach, the moments M_{cxn} and M_{cyn} are calculated using lever arms Lax_{mrc} and Lay_{mrc}) measured relative to the original, unrotated reference axes. In the second method, involving inclined or rotated moment reference axes, the moments M_{cxi} and M_{cyi} are determined using lever arms Lax_{mrci} and Lay_{mrci}, which are calculated relative to axes that are inclined and perpendicular to the neutral axis.

4.3.1 Method 1: Non-inclined (Unrotated) Axis Method

The method utilizes the non-inclined moment reference axis, followed by worked example calculations as follows.

4.3.1.1 Method of Calculation

The resistive bending moment in concrete along the non-inclined x-axis (M_{cxn}) and non-inclined y-axis (M_{cyn}) are determined utilizing the following approach. This is shown in Equations 4.3 and 4.4, and illustrated in Figure 4.1.

The Delta (δ) and Lambda (λ) are Greek letters utilized for notation.

$$M_{cxn} = N_c \cdot Lax_{mrc} \tag{4.3}$$

$$M_{cyn} = N_c \cdot Lay_{mrc} \tag{4.4}$$

Where:

M_{cxn}: Resistive bending moment in concrete along the non-inclined x-axis.

M_{cyn}: Resistive bending moment in concrete along the non-inclined y-axis.

N_c: Resistive axial compressive force in concrete.

Lax_{mrc}: Lever arm for concrete from non-inclined moment reference axis x to the centroid of compression zone.

Lay_{mrc}: Lever arm for concrete from non-inclined moment reference axis y to the centroid of compression zone.

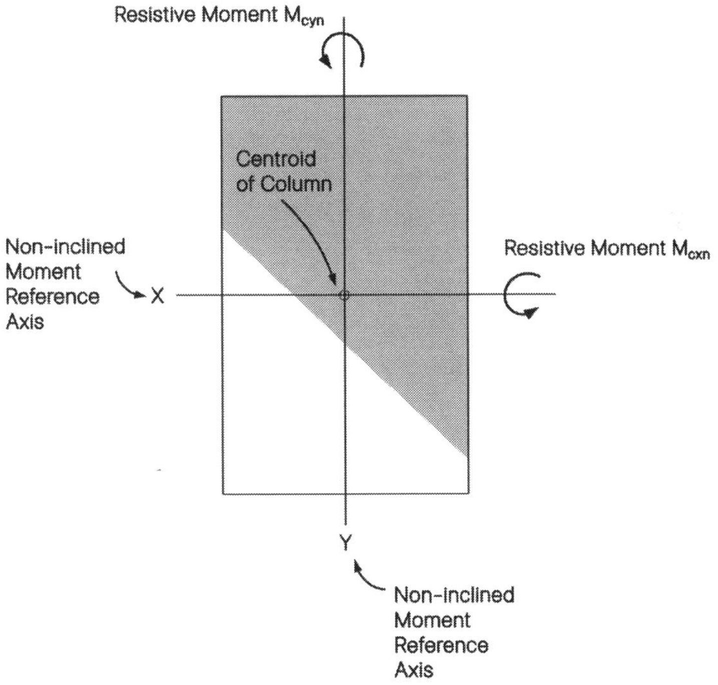

Figure 4.1: Resistive moments along the non-inclined x and y axis

4.3.1.2 Example Calculation of Moments (M_{cxn} and M_{cyn})

The moment calculation examples for each behaviour case are as follows.

Behaviour Case 1: $\alpha\delta b \leq b$, **and** $\alpha\delta h \leq h$

Given Data: $N_c = 235008\,\text{N}$, $Lax_{mrc} = 96.667\,\text{mm}$, $Lay_{mrc} = 48.5\,\text{mm}$.

Calculation:

$$\begin{aligned}
M_{cxn} &= N_c \cdot Lax_{mrc} \\
&= 235008\,\text{N} \times 96.667\,\text{mm} \\
&= 22717518.336\,\text{Nmm} \\
&= 22717518.336\,\text{Nmm} \times 10^{-6} = 22.718\,\text{kNm}
\end{aligned}$$

$$\begin{aligned}
M_{cyn} &= N_c \cdot Lay_{mrc} \\
&= 235008\,\text{N} \times 48.5\,\text{mm} \\
&= 11397888\,\text{Nmm} \\
&= 11397888\,\text{Nmm} \times 10^{-6} = 11.398\,\text{kNm}
\end{aligned}$$

Behaviour Case 2: $\alpha\delta b > b$, **and** $\alpha\delta h \leq h$

Given Data: $N_c = 172757.808\,\text{N}$, $Lax_{mrc} = 122.863\,\text{mm}$, $Lay_{mrc} = 18.544\,\text{mm}$.

Calculation:

$$\begin{aligned}
M_{cxn} &= N_c \cdot Lax_{mrc} \\
&= 172757.808\,\text{N} \times 122.683\,\text{mm} \\
&= 21194446.159\,\text{Nmm} \\
&= 21194446.159\,\text{Nmm} \times 10^{-6} = 21.194\,\text{kNm}
\end{aligned}$$

$$M_{cyn} = N_c \cdot Lay_{mrc}$$

$$
\begin{aligned}
&= 172757.808\,\mathrm{N} \times 18.544\,\mathrm{mm} \\
&= 3203620.791\,\mathrm{Nmm} \\
&= 3203620.791\,\mathrm{Nmm} \times 10^{-6} = 3.204\,\mathrm{kNm}
\end{aligned}
$$

Behaviour Case 3: $\alpha\delta b \leq b$**, and** $\alpha\delta h > h$

Given Data: $N_c = 468180\,\mathrm{N}$, $Lax_{mrc} = 62.500\,\mathrm{mm}$, $Lay_{mrc} = 53.944\,\mathrm{mm}$.

Calculation:

$$
\begin{aligned}
M_{cxn} &= N_c \cdot Lax_{mrc} \\
&= 468180\,\mathrm{N} \times 62.500\,\mathrm{mm} \\
&= 29261250\,\mathrm{Nmm} \\
&= 29261250\,\mathrm{Nmm} \times 10^{-6} = 29.261\,\mathrm{kNm}
\end{aligned}
$$

$$
\begin{aligned}
M_{cyn} &= N_c \cdot Lay_{mrc} \\
&= 468180\,\mathrm{N} \times 53.944\,\mathrm{mm} \\
&= 25255501.92\,\mathrm{Nmm} \\
&= 25255501.92\,\mathrm{Nmm} \times 10^{-6} = 25.256\,\mathrm{kNm}
\end{aligned}
$$

Behaviour Case 4: $\alpha\delta b > b$**, and** $\alpha\delta h > h$

Given Data: $N_c = 824773.778\,\mathrm{N}$, $Lax_{mrc} = 23.269\,\mathrm{mm}$, $Lay_{mrc} = 15.171\,\mathrm{mm}$.

Calculation:

$$
\begin{aligned}
M_{cxn} &= N_c \cdot Lax_{mrc} \\
&= 824773.778\,\mathrm{N} \times 23.269\,\mathrm{mm} \\
&= 19191661.040\,\mathrm{Nmm} \\
&= 19191661.040\,\mathrm{Nmm} \times 10^{-6} = 19.192\,\mathrm{kNm}
\end{aligned}
$$

$$
\begin{aligned}
M_{cyn} &= N_c \cdot Lay_{mrc} \\
&= 824773.778\,\mathrm{N} \times 15.171\,\mathrm{mm}
\end{aligned}
$$

$$= 12512642.986\,\text{Nmm}$$
$$= 12512642.986\,\text{Nmm} \times 10^{-6} = 12.513\,\text{kNm}$$

4.3.1.3 Summary of Calculations

The following Table 4.1 shows the summary of moments for each inclined axis M_{cxn} and M_{cyn}.

Table 4.1: Summary of the moments for each behaviour case

Moment Type	Case 1	Case 2	Case 3	Case 4
M_{cxn}	22.718 kNm	21.194 kNm	29.261 kNm	19.192 kNm
M_{cyn}	11.398 kNm	3.204 kNm	25.256 kNm	12.513 kNm

4.4 Method 2: Inclined (Rotated) Axis Method

The method utilizes the inclined moment reference axis, followed by example calculations as follows.

4.4.1 Method of Calculation

The resistive bending moment in concrete along the inclined x-axis (M_{cxi}) and inclined y-axis (M_{cyi}) are determined using the following approach. This is shown in Equations 4.5 and 4.6, and illustrated in Figure 4.2. The Delta (δ) and Lambda (λ) are Greek letters utilized for notation.

$$M_{cxi} = N_c \cdot Lax_{mrci} \tag{4.5}$$

$$M_{cyi} = N_c \cdot Lay_{mrci} \tag{4.6}$$

Where:

M_{cxi}: Resistive bending moment in concrete along the inclined x-axis.
M_{cyi}: Resistive bending moment in concrete along the inclined y-axis.
N_c: Resistive axial compressive force in concrete.

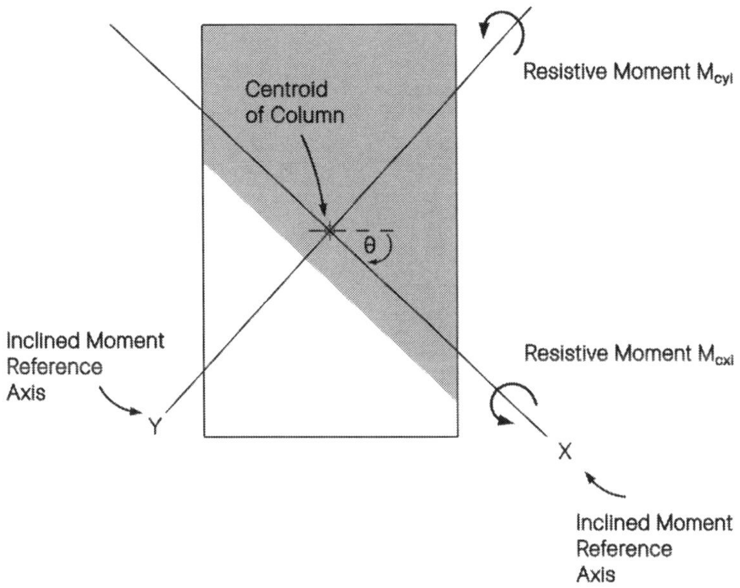

Figure 4.2: Resistive moments along the inclined x and y axis

Lax_{mrci}: Lever arm for concrete from inclined moment reference axis x to the centroid of compression zone.
Lay_{mrci}: Lever arm for concrete from inclined moment reference axis y to the centroid of compression zone.

4.4.2 Example Calculation of Moments (M_{cxi} and M_{cyi})

The moment calculation examples for each behaviour case are as follows.

Behaviour Case 1: $\alpha\delta b \leq b$, **and** $\alpha\delta h \leq h$

Given Data: $N_c = 235008\,\text{N}$, $Lax_{mrc} = 96.667\,\text{mm}$, $Lay_{mrc} = 48.5\,\text{mm}$, $\theta = 39.805571092°$.

Calculation:

$$
\begin{aligned}
Lax_{mrci} &= (Lax_{mrc} \cdot \cos(\theta)) + (Lay_{mrc} \cdot \sin(\theta)) \\
&= (96.667\,\text{mm} \times \cos(39.805571092^\circ)) + (48.5\,\text{mm} \times \sin(39.805571092^\circ)) \\
&= +105.31\,\text{mm}
\end{aligned}
$$

$$
\begin{aligned}
Lay_{mrci} &= (Lay_{mrc} \cdot \cos(\theta)) - (Lax_{mrc} \cdot \sin(\theta)) \\
&= (48.5\,\text{mm} \times \cos(39.805571092^\circ)) - (96.667\,\text{mm} \times \sin(39.805571092^\circ)) \\
&= -24.63\,\text{mm}
\end{aligned}
$$

$$
\begin{aligned}
M_{cxi} &= N_c \cdot Lax_{mrci} \\
&= 235008\,\text{N} \times 105.31\,\text{mm} \\
&= 24748692.48\,\text{Nmm} \\
&= 24748692.48\,\text{Nmm} \times 10^{-6} = 24.749\,\text{kNm}
\end{aligned}
$$

$$
\begin{aligned}
M_{cyi} &= N_c \cdot Lay_{mrci} \\
&= 235008\,\text{N} \times (-24.63\,\text{mm}) \\
&= -5788247.04\,\text{Nmm} \\
&= -5788247.04\,\text{Nmm} \times 10^{-6} = -5.788\,\text{kNm}
\end{aligned}
$$

Behaviour Case 2: $\alpha\delta b > b$, and $\alpha\delta h \leq h$

Given Data: $N_c = 172757.808\,\text{N}$, $Lax_{mrc} = 122.863\,\text{mm}$, $Lay_{mrc} = 18.544\,\text{mm}$, $\theta = 39.805571092^\circ$.

Calculation:

$$
\begin{aligned}
Lax_{mrci} &= (Lax_{mrc} \cdot \cos(\theta)) + (Lay_{mrc} \cdot \sin(\theta)) \\
&= (122.863\,\text{mm} \times \cos(39.805571092^\circ)) + (18.544\,\text{mm} \\
&\quad \times \sin(39.805571092^\circ)) \\
&= +106.26\,\text{mm}
\end{aligned}
$$

$$
\begin{aligned}
Lay_{mrci} &= (Lay_{mrc} \cdot \cos(\theta)) - (Lax_{mrc} \cdot \sin(\theta)) \\
&= (18.544\,\text{mm} \times \cos(39.805571092^\circ)) - (122.863\,\text{mm} \\
&\quad \times \sin(39.805571092^\circ))
\end{aligned}
$$

$$= -64.41\,\text{mm}$$

$$\begin{aligned}M_{cxi} &= N_c \cdot Lax_{mrci}\\ &= 172757.808\,\text{N} \times 106.26\,\text{mm}\\ &= 18357244.68\,\text{Nmm}\\ &= 18357244.68\,\text{Nmm} \times 10^{-6} = 18.36\,\text{kNm}\end{aligned}$$

$$\begin{aligned}M_{cyi} &= N_c \cdot Lay_{mrci}\\ &= 172757.808\,\text{N} \times (-64.41\,\text{mm})\\ &= -11127330.41\,\text{Nmm}\\ &= -11127330.41\,\text{Nmm} \times 10^{-6} = -11.13\,\text{kNm}\end{aligned}$$

Behaviour Case 3: $\alpha\delta b \leq b$, and $\alpha\delta h > h$

Given Data: $N_c = 468180\,\text{N}$, $Lax_{mrc} = 62.500\,\text{mm}$, $Lay_{mrc} = 53.944\,\text{mm}$, $\theta = 39.805571092°$.

Calculation:

$$\begin{aligned}Lax_{mrci} &= (Lax_{mrc} \cdot \cos(\theta)) + (Lay_{mrc} \cdot \sin(\theta))\\ &= (62.500\,\text{mm}\\ &\quad \times \cos(39.805571092°)) + (53.944\,\text{mm} \times \sin(39.805571092°))\\ &= +82.55\,\text{mm}\end{aligned}$$

$$\begin{aligned}Lay_{mrci} &= (Lay_{mrc} \cdot \cos(\theta)) - (Lax_{mrc} \cdot \sin(\theta))\\ &= (53.944\,\text{mm}\\ &\quad \times \cos(39.805571092°)) - (62.500\,\text{mm} \times \sin(39.805571092°))\\ &= +1.43\,\text{mm}\end{aligned}$$

$$\begin{aligned}M_{cxi} &= N_c \cdot Lax_{mrci}\\ &= 468180\,\text{N} \times 82.55\,\text{mm}\\ &= 38648259\,\text{Nmm}\\ &= 38648259\,\text{Nmm} \times 10^{-6} = 38.648\,\text{kNm}\end{aligned}$$

$$M_{cyi} = N_c \cdot Lay_{mrci}$$

$$= 468180\,\text{N} \times 1.43\,\text{mm}$$
$$= 669497.4\,\text{Nmm}$$
$$= 669497.4\,\text{Nmm} \times 10^{-6} = 0.669\,\text{kNm}$$

Behaviour Case 4: $\alpha\delta b > b$**, and** $\alpha\delta h > h$

Given Data: $N_c = 824773.780\,\text{N}$, $Lax_{mrc} = 23.269\,\text{mm}$, $Lay_{mrc} = 15.171\,\text{mm}$, $\theta = 39.805571092°$.

Calculation:

$$\begin{aligned}
Lax_{mrci} &= (Lax_{mrc} \cdot \cos(\theta)) + (Lay_{mrc} \cdot \sin(\theta)) \\
&= (23.269\,\text{mm} \times \cos(39.805571092°)) + (15.171\,\text{mm} \\
&\quad \times \sin(39.805571092°)) \\
&= +27.59\,\text{mm}
\end{aligned}$$

$$\begin{aligned}
Lay_{mrci} &= (Lay_{mrc} \cdot \cos(\theta)) - (Lax_{mrc} \cdot \sin(\theta)) \\
&= (15.171\,\text{mm} \times \cos(39.805571092°)) - (23.269\,\text{mm} \\
&\quad \times \sin(39.805571092°)) \\
&= -3.24\,\text{mm}
\end{aligned}$$

$$\begin{aligned}
M_{cxi} &= N_c \cdot Lax_{mrci} \\
&= 824773.780\,\text{N} \times 27.59\,\text{mm} \\
&= 22755508.59\,\text{Nmm} \\
&= 22755508.59\,\text{Nmm} \times 10^{-6} = 22.756\,\text{kNm}
\end{aligned}$$

$$\begin{aligned}
M_{cyi} &= N_c \cdot Lay_{mrci} \\
&= 824773.780\,\text{N} \times (-3.24\,\text{mm}) \\
&= -2672267.047\,\text{Nmm} \\
&= -2672267.047\,\text{Nmm} \times 10^{-6} = -2.672\,\text{kNm}
\end{aligned}$$

4.4.3 Summary of Calculations

The following Table 4.2 shows the summary of moments for each inclined axis M_{cxi} and M_{cyi}.

Table 4.2: Summary of the moments for each behaviour case

Moment Type	**Case 1**	**Case 2**	**Case 3**	**Case 4**
M_{cxi}	24.75 kNm	18.36 kNm	38.648 kNm	22.756 kNm
M_{cyi}	-5.79 kNm	-11.13 kNm	0.669 kNm	-2.672 kNm

This page intentionally left blank.

Chapter 5:

The Lever Arms for Tensile and Compressive Steel

This chapter describes the lever arms utilized to calculate resistive axial force and moments in steel.

5.1 The Lever Arms Lax_{cfs} and Lay_{cfs}

Figure 5.1 and 5.2 shows the lever arm distances from the extreme compression fibre to the centroid of the reinforcement.

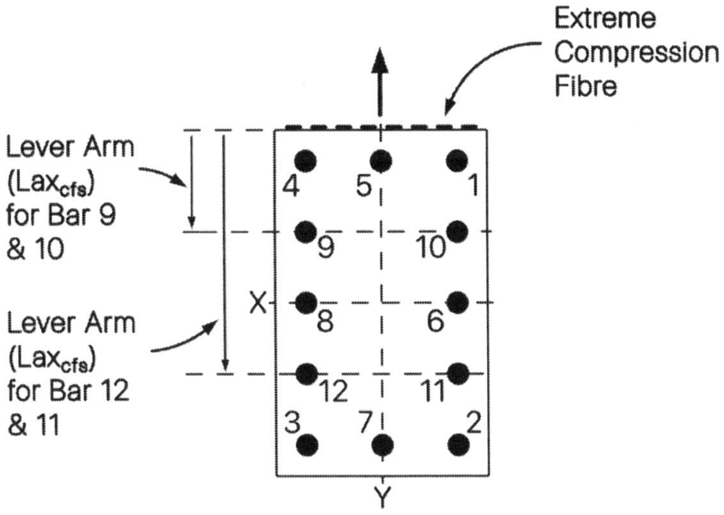

Figure 5.1: The lever arm Lax_{cfs}

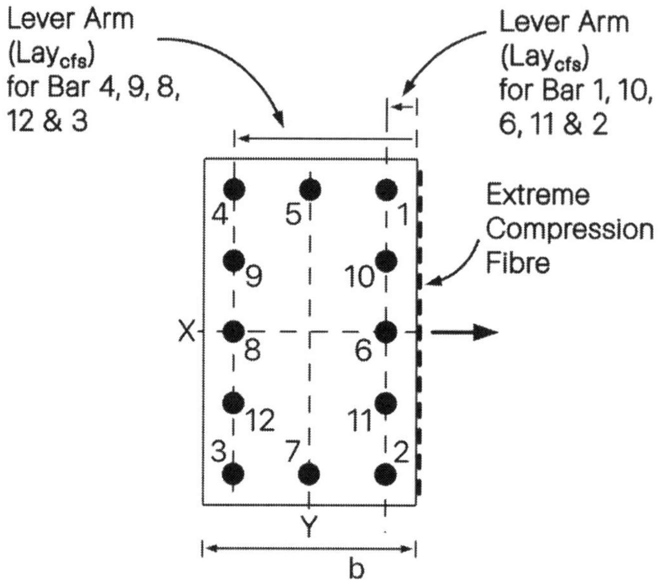

Figure 5.2: The lever arm Lay_{cfs}

5.2 The Lever Arms Lax_{mrs} and Lay_{mrs}

The lever arms Lax_{mrs} and Lay_{mrs} are simply the distance from the moment reference axis x or y, to the centroid of each reinforcement bar.

The centroid of a reinforcement bar refers to its geometric centre, the point at which the mass of the bar is considered to be concentrated. When calculating lever arms, perpendicular distance is measured from the moment reference axis to the centroid of each bar. This distance helps understand how much each bar contributes to resisting the bending moments about the x and y axes.

5.2.1 Method of Calculation

Each distance is utilized to determine the contribution of the reinforcement bars to the overall bending resistance in the respective direction, x or y. Moreover, bars 6 and 8 for the x-axis and bars 5 and 7 for the y-axis lie exactly on the moment reference axes, thereby providing no lever arm distance for those reinforcement bars. This means the lever arm should be taken as zero. The lever arm distances in the compression zone are considered positive (+), and the lever arm distances in the tensile zone are considered negative (-). This is illustrated in Figures 5.3 and 5.4.

Although the examples throughout this book utilize only four reinforcement bars, Figures 5.3 and 5.4 shows twelve bars. This is intended to demonstrate the method of calculating the lever arms when more reinforcement bars are involved. Moreover, this illustrates the concept of zero lever arm distance for bars that lie directly on the moment reference axes. The methods outlined in this book are universally applicable to any number of reinforcement bars and to layouts of any size or configuration.

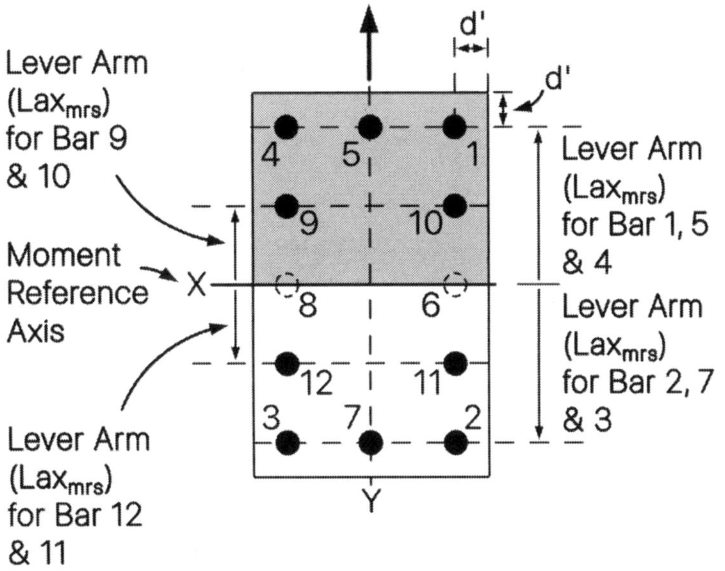

Figure 5.3: The lever arm Lax_{mrs}

The Equation 5.1 shows the calculation formula for the lever arm Lax_{mrs} along the x-axis, while Equation 5.2 shows the calculation formula for the lever arm Lay_{mrs} along the y-axis.

$$Lax_{mrs} = 0.5h - Lax_{cfs} \tag{5.1}$$

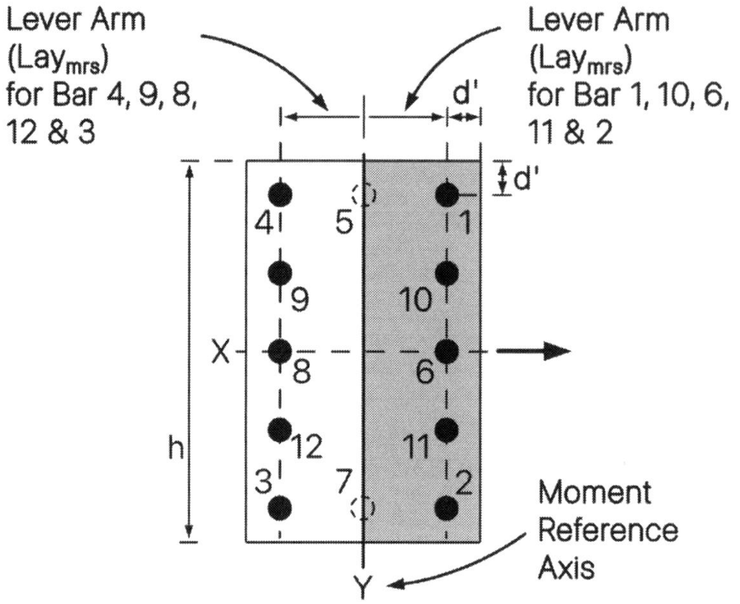

Figure 5.4: The lever arm Lay_{mrs}

$$Lay_{mrs} = 0.5b - Lay_{cfs} \tag{5.2}$$

Where,

Lax_{mrs}: Lever arm for steel from moment reference axis x to the centroid of reinforcement bar.

Lay_{mrs}: Lever arm for steel from moment reference axis y to the centroid of reinforcement bar.

Lax_{cfs}: Lever arm for steel for x axis from extreme compression fibre to the centroid of reinforcement bar.

Lay_{cfs}: Lever arm for steel for y axis from extreme compression fibre to the centroid of reinforcement bar.

h: Height of column.

b: Breadth of column.

5.2.2 Example Calculation

Given Data:

$h = 300\,\text{mm},\ b = 225\,\text{mm},\ \delta h = 200\,\text{mm},\ \delta b = 240\,\text{mm},\ \theta = 39.805571092°$

Lever Arm	Bar 1	Bar 2	Bar 3	Bar 4
Lax_{cfs}	40 mm	260 mm	260 mm	40 mm
Lay_{cfs}	40 mm	40 mm	185 mm	185 mm

Calculation:

$$Lax_{mrs} = (0.5 \cdot h) - Lax_{cfs}$$

$$\begin{aligned}
\text{Bar 1:}\quad & 0.5 \times 300\,\text{mm} - 40\,\text{mm} = 150\text{mm} - 40\,\text{mm} = 110\,\text{mm} \\
\text{Bar 2:}\quad & 0.5 \times 300\,\text{mm} - 260\,\text{mm} = 150\,\text{mm} - 260\,\text{mm} = -110\,\text{mm} \\
\text{Bar 3:}\quad & 0.5 \times 300\,\text{mm} - 260\,\text{mm} = 150\,\text{mm} - 260\,\text{mm} = -110\,\text{mm} \\
\text{Bar 4:}\quad & 0.5 \times 300\,\text{mm} - 40\,\text{mm} = 150\,\text{mm} - 40\,\text{mm} = 110\,\text{mm}
\end{aligned}$$

$$Lay_{mrs} = (0.5 \cdot b) - Lay_{cfs}$$

$$\begin{aligned}
\text{Bar 1:}\quad & 0.5 \times 225\,\text{mm} - 40\,\text{mm} = 112.5\,\text{mm} - 40\,\text{mm} = 72.5\,\text{mm} \\
\text{Bar 2:}\quad & 0.5 \times 225\,\text{mm} - 40\,\text{mm} = 112.5\,\text{mm} - 40\,\text{mm} = 72.5\,\text{mm} \\
\text{Bar 3:}\quad & 0.5 \times 225\,\text{mm} - 185\,\text{mm} = 112.5\,\text{mm} - 185\,\text{mm} = -72.5\,\text{mm} \\
\text{Bar 4:}\quad & 0.5 \times 225\,\text{mm} - 185\,\text{mm} = 112.5\,\text{mm} - 185\,\text{mm} = -72.5\,\text{mm}
\end{aligned}$$

5.2.3 Summary of Calculations

Lever Arm	Bar 1	Bar 2	Bar 3	Bar 4
Lax_{mrs}	+110 mm	-110 mm	-110 mm	+110 mm
Lay_{mrs}	+72.5 mm	+72.5 mm	-72.5 mm	-72.5 mm

5.3 The Lever Arms Lax_{mrsi} and Lay_{mrsi}

Distances from reinforcement centroids to the inclined (rotated) moment reference axes (x and y axes) are determined by accounting for the inclination angle (θ) of the axis relative to the unrotated x axis.

The rotated axes intersect at the centroid of the section, ensuring consistency with the actual geometry of the reinforced section.

The calculations consider two approaches,

Method 1: This uses the direct projection of horizontal and vertical distances onto the inclined axes.

Method 2: This applies geometric relationships to determine the distances considering the inclination of the axes.

This is illustrated in Figure 5.5.

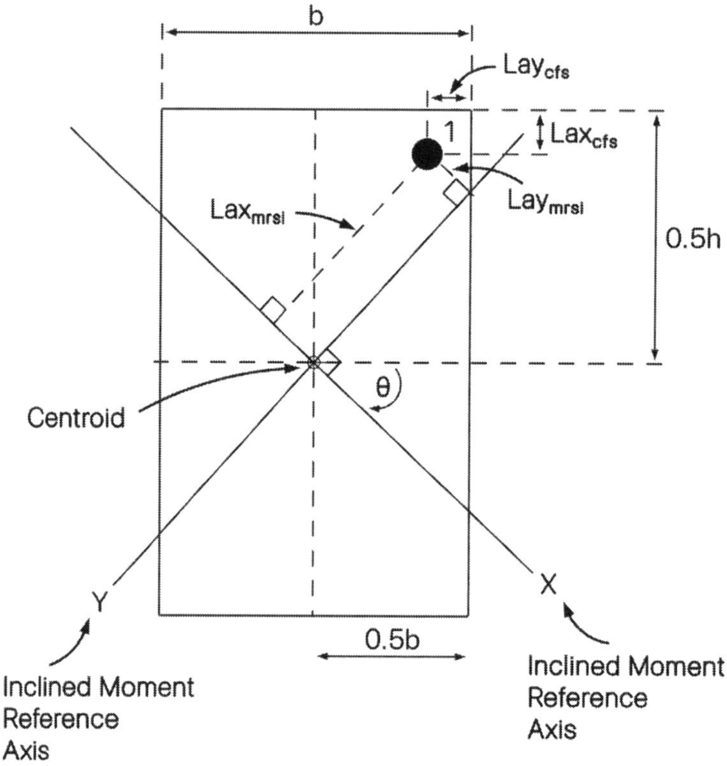

Figure 5.5: The lever arms Lax_{mrsi} and Lay_{mrsi} for method 1

5.3.1 Method 1

In this method, the distances from the reinforcement centroids to the inclined moment reference axes (rotated x and y axes) by directly projecting the horizontal Lax_{mrs} and vertical Lay_{mrs} distances onto the inclined axes. The projections are calculated using trigonometric relationships, taking into account the angle of inclination (θ) of the axes relative to the unrotated x axis. This method aligns with the geometric orientation of the reinforced section and ensures consistency in determining the lever arms. This is shown in the Equations 5.3 and 5.4.

5.3.1.1 Method of Calculation

$$Lax_{mrsi} = (Lax_{mrs} \cdot \cos(\theta)) + (Lay_{mrs} \cdot \sin(\theta)) \tag{5.3}$$

$$Lay_{mrsi} = (Lay_{mrs} \cdot \cos(\theta)) - (Lax_{mrs} \cdot \sin(\theta)) \tag{5.4}$$

Where,
Lax_{mrsi}: Distance from reinforcement centroid to inclined (rotated) moment reference axis x.
Lay_{mrsi}: Distance from reinforcement centroid to inclined (rotated) moment reference axis y.

5.3.1.2 Example Calculation

Given Data:

$h = 300$ mm, $b = 225$ mm, $\delta h = 200$ mm, $\delta b = 240$ mm, $\theta = 39.805571092°$

Bar Number	Lax_{mrs}	Lay_{mrs}
Bar 1	+110 mm	+72.5 mm
Bar 2	-110 mm	+72.5 mm
Bar 3	-110 mm	-72.5 mm
Bar 4	+110 mm	-72.5 mm

Calculation:

$$Lax_{mrsi} = (Lax_{mrs} \cdot \cos(\theta)) + (Lay_{mrs} \cdot \sin(\theta))$$
$$Lay_{mrsi} = (Lay_{mrs} \cdot \cos(\theta)) - (Lax_{mrs} \cdot \sin(\theta))$$

$$\cos(\theta) = \cos(39.805571092^\circ) = 0.76822128$$
$$\sin(\theta) = \sin(39.805571092^\circ) = 0.6401844$$

Bar 1:

$$\begin{aligned} Lax_{mrsi} &= (110\,\text{mm} \times 0.76822128) + (72.5\,\text{mm} \times 0.6401844) \\ &= 130.918\,\text{mm} \end{aligned}$$

$$\begin{aligned} Lay_{mrsi} &= (72.5\,\text{mm} \times 0.76822128) - (110\,\text{mm} \times 0.6401844) \\ &= -14.724\,\text{mm} \end{aligned}$$

Bar 2:

$$\begin{aligned} Lax_{mrsi} &= (-110\,\text{mm} \times 0.76822128) + (72.5\,\text{mm} \times 0.6401844) \\ &= -38.091\,\text{mm} \end{aligned}$$

$$\begin{aligned} Lay_{mrsi} &= (72.5\,\text{mm} \times 0.76822128) - (-110\,\text{mm} \times 0.6401844) \\ &= 126.116\,\text{mm} \end{aligned}$$

Bar 3:

$$\begin{aligned} Lax_{mrsi} &= (-110\,\text{mm} \times 0.76822128) + (-72.5\,\text{mm} \times 0.6401844) \\ &= -130.918\,\text{mm} \end{aligned}$$

$$\begin{aligned} Lay_{mrsi} &= (-72.5\,\text{mm} \times 0.76822128) - (-110\,\text{mm} \times 0.6401844) \\ &= 14.724\,\text{mm} \end{aligned}$$

Bar 4:

$$Lax_{mrsi} = (110\,\text{mm} \times 0.76822128) + (-72.5\,\text{mm} \times 0.6401844)$$

$$= 38.091\,\text{mm}$$

$$Lay_{mrsi} = (-72.5\,\text{mm} \times 0.76822128) - (110\,\text{mm} \times 0.6401844)$$
$$= -126.116\,\text{mm}$$

5.3.2 Method 2

This method employs geometric principles to calculate distances directly from reinforcement centroids to the inclined (rotated) axes. By aligning the rotated axes with the neutral axis inclination (θ), distances are calculated through sequential steps involving projections and trigonometric relationships.

This is divided into two main sections,

1. Distance from inclined (rotated) x axis (Lax_{mrsi})
2. Distance from inclined (rotated) y axis (Lay_{mrsi})

The utilization of basic geometry to find the distances from each rotated axis is detailed as follows.

5.3.2.1 Distance from Inclined (Rotated) x Axis (Lax_{mrsi})

This section calculates the horizontal distances Lax_{mrsi} from reinforcement centroids to the inclined (rotated) x-axis. Using basic geometry, the distances are derived by projecting reinforcement centroid coordinates onto the inclined axis, considering the inclination angle (θ) and the geometric relationships within the section.

5.3.2.1.1 Method of Calculation

The following shows the basic method to calculate the distance from reinforcement centroid to inclined (rotated) x axis. Moreover, the Equations 5.5 to 5.9 show the final derived equations for each calculation step. This is illustrated in Figure 5.6.

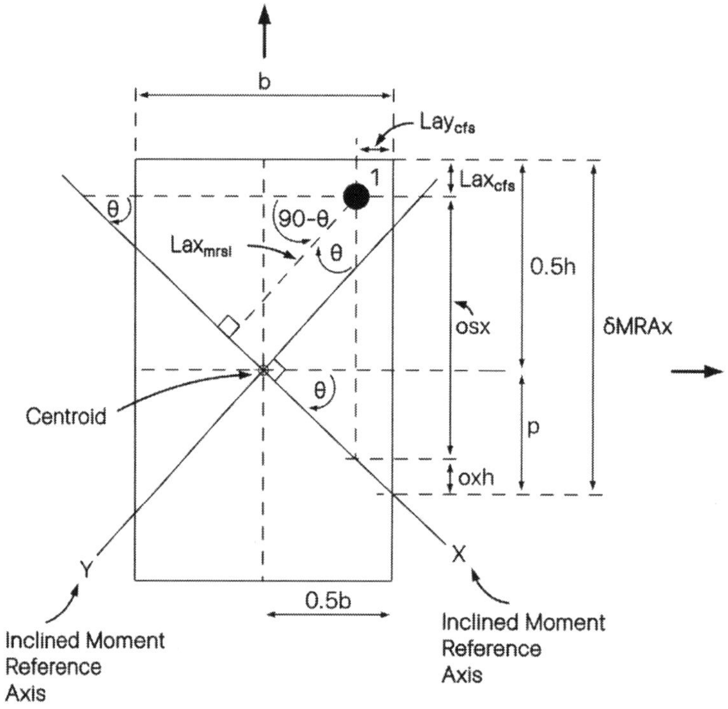

Figure 5.6: The lever arm Lax_{mrsi} for method 2

Derivation of Lax_{mrsi}:

$$\cos(\theta) = \frac{\text{adjacent}}{\text{hypotenuse}}$$

Thereby,

$$\cos(\theta) = \frac{Lax_{mrsi}}{\text{osx}}$$

$$Lax_{mrsi} = \cos(\theta) \cdot \text{osx}$$

Derivation of osx:

$$\text{osx} = \delta MRAx - (\text{oxh} + Lax_{cfs})$$

Derivation of oxh:

$$\tan(\theta) = \frac{\text{opposite}}{\text{adjacent}}$$

Threrefore,

$$\tan(\theta) = \frac{\text{oxh}}{Lay_{cfs}}$$
$$\text{oxh} = \tan(\theta) \cdot Lay_{cfs}$$

Derivation of $\delta MRAx$:

$$\delta MRAx = (0.5 \cdot h) + P$$

Derivation of P:

$$\tan(\theta) = \frac{\text{opposite}}{\text{adjacent}}$$
$$\tan(\theta) = \frac{P}{0.5 \cdot b}$$
$$P = \tan(\theta) \cdot (0.5 \cdot b)$$

Summary of derived equations

$$Lax_{mrsi} = \cos(\theta) \cdot osx \tag{5.5}$$

$$osx = \delta MRAx - (oxh + Lax_{cfs}) \tag{5.6}$$

$$oxh = \tan(\theta) \cdot Lay_{cfs} \tag{5.7}$$

$$\delta MRAx = (0.5 \cdot h) + P \tag{5.8}$$

$$P = \tan(\theta) \cdot (0.5 \cdot b) \tag{5.9}$$

Where,

Lax_{mrsi}: Distance from reinforcement centroid to inclined (rotated) moment reference axis x.
Lax_{cfs}: Lever arm for steel for x axis from extreme compression fibre to the centroid of reinforcement bar.
Lay_{cfs}: Lever arm for steel for y axis from extreme compression fibre to the centroid of reinforcement bar.
h: Height of column.
b: Breadth of column.
δh: Neutral axis distance along the height.
δb: Neutral axis distance along the breadth.

5.3.2.1.2 Example Calculations

Given Data:

$h = 300\,\text{mm}$, $b = 225\,\text{mm}$, $\delta h = 200\,\text{mm}$, $\delta b = 240\,\text{mm}$.

Lever Arm	Bar 1	Bar 2	Bar 3	Bar 4
Lax_{cfs}	40 mm	260 mm	260 mm	40 mm
Lay_{cfs}	40 mm	40 mm	185 mm	185 mm

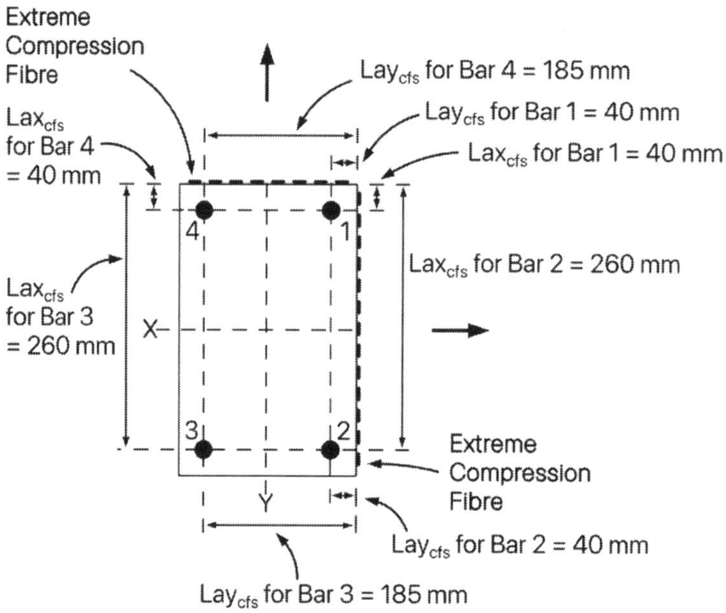

Figure 5.7: The lever arms Lax_{cfs} and Lay_{cfs}

Calculation of Inclination Angle (θ):

$$\text{Inclination Ratio (IR)} = \frac{\delta h}{\delta b} = \frac{200\,\text{mm}}{240\,\text{mm}} = 0.833333333$$

$$\text{Inclination Angle } (\theta) = \arctan(\text{IR}) = \arctan(0.833333333) = 39.805571092^\circ$$

Note: When using spreadsheet software, ensure that the output of the *arctan* function is converted to degrees by using DEGREES(ATAN(IR)), as some software returns ATAN values in radians by default.

Calculation of P:

$$\begin{aligned} P &= \tan(\theta) \cdot (0.5 \cdot b) \\ &= \tan(39.80557109^\circ) \times (0.5 \times 225\,\text{mm}) \\ &= 0.833333333 \times 112.5\,\text{mm} \\ &= 93.75\,\text{mm} \end{aligned}$$

Note: The inclination ratio (IR) is equivalent to $\tan(\theta)$, so it can directly be used in place of $\tan(\theta)$ for simplified calculations.

Calculation of $\delta MRAx$:

$$\begin{aligned} \delta MRAx &= (0.5 \cdot h) + P \\ &= (0.5 \times 300\,\text{mm}) + 93.75\,\text{mm} \\ &= 243.75\,\text{mm} \end{aligned}$$

Bar 1:

$$\begin{aligned} \text{oxh} &= \tan(\theta) \cdot Lay_{cfs} \\ &= \tan(39.805571092^\circ) \times 40\,\text{mm} \\ &= 33.333\,\text{mm} \end{aligned}$$

$$\begin{aligned} \text{osx} &= \delta MRAx - (\text{oxh} + Lax_{cfs}) \\ &= 243.75\,\text{mm} - (33.333\,\text{mm} + 40\,\text{mm}) \\ &= 170.417\,\text{mm} \end{aligned}$$

$$\begin{aligned} Lax_{mrsi} &= \cos(\theta) \cdot \text{osx} \\ &= \cos(39.80557109^\circ) \times 170.417\,\text{mm} \\ &= 130.918\,\text{mm} \end{aligned}$$

Bar 2:

$$\begin{aligned}
\text{oxh} &= \tan(\theta) \cdot Lay_{cfs} \\
&= \tan(39.805571092^\circ) \times 40\,\text{mm} \\
&= 33.333\,\text{mm}
\end{aligned}$$

$$\begin{aligned}
\text{osx} &= \delta MRAx - (\text{oxh} + Lax_{cfs}) \\
&= 243.75\,\text{mm} - (33.333\,\text{mm} + 260\,\text{mm}) \\
&= -49.583\,\text{mm}
\end{aligned}$$

$$\begin{aligned}
Lax_{mrsi} &= \cos(\theta) \cdot \text{osx} \\
&= \cos(39.805571092^\circ) \times -49.583\,\text{mm} \\
&= -38.091\,\text{mm}
\end{aligned}$$

Bar 3:

$$\begin{aligned}
\text{oxh} &= \tan(\theta) \cdot Lay_{cfs} \\
&= \tan(39.805571092^\circ) \times 185\,\text{mm} \\
&= 154.167\,\text{mm}
\end{aligned}$$

$$\begin{aligned}
\text{osx} &= \delta MRAx - (\text{oxh} + Lax_{cfs}) \\
&= 243.75\,\text{mm} - (154.167\,\text{mm} + 260\,\text{mm}) \\
&= -170.417\,\text{mm}
\end{aligned}$$

$$\begin{aligned}
Lax_{mrsi} &= \cos(\theta) \cdot \text{osx} \\
&= \cos(39.805571092^\circ) \times -170.417\,\text{mm} \\
&= -130.918\,\text{mm}
\end{aligned}$$

Bar 4:

$$\begin{aligned}
\text{oxh} &= \tan(\theta) \cdot Lay_{cfs} \\
&= \tan(39.805571092^\circ) \times 185\,\text{mm} \\
&= 154.167\,\text{mm}
\end{aligned}$$

$$\text{osx} = \delta MRAx - (\text{oxh} + Lax_{cfs})$$

$$
\begin{aligned}
&= 243.75\,\text{mm} - (154.167\,\text{mm} + 40\,\text{mm}) \\
&= 49.583\,\text{mm}
\end{aligned}
$$

$$
\begin{aligned}
Lax_{mrsi} &= \cos(\theta) \cdot \text{osx} \\
&= \cos(39.805571092°) \times 49.583\,\text{mm} \\
&= 38.091\,\text{mm}
\end{aligned}
$$

5.3.2.2 Distance from Inclined (Rotated) y Axis (Lay_{mrsi})

This section calculates the horizontal distances Lay_{mrsi} from reinforcement centroids to the inclined (rotated) y-axis. Using basic geometry, the distances are derived by projecting reinforcement centroid coordinates onto the inclined axis, considering the inclination angle (θ) and the geometric relationships within the section.

5.3.2.2.1 Method of Calculation

The following shows the basic method of calculation for the distance from the reinforcement centroid to the inclined (rotated) y axis. Moreover, the Equations 5.10 to 5.14 show the final derived equations for each calculation step, and this is illustrated in Figures 5.8 to 5.11. Figures 5.8 and 5.9 show a case where the inclined y axis does not surpass bar 1. Figures 5.10 and 5.11 show a case where the inclined y axis surpasses bar 1. Both cases use the same method to calculate Lay_{mrsi}.

Derivation of Lay_{mrsi}:

$$
\begin{aligned}
\cos(\theta) &= \frac{\text{adjacent}}{\text{hypotenuse}} = \frac{Lay_{mrsi}}{\text{osy}} \\
Lay_{mrsi} &= \cos(\theta) \cdot \text{osy}
\end{aligned}
$$

Derivation of osy:

$$
\text{osy} = \delta MRAy - (\text{oyb} + Lay_{cfs})
$$

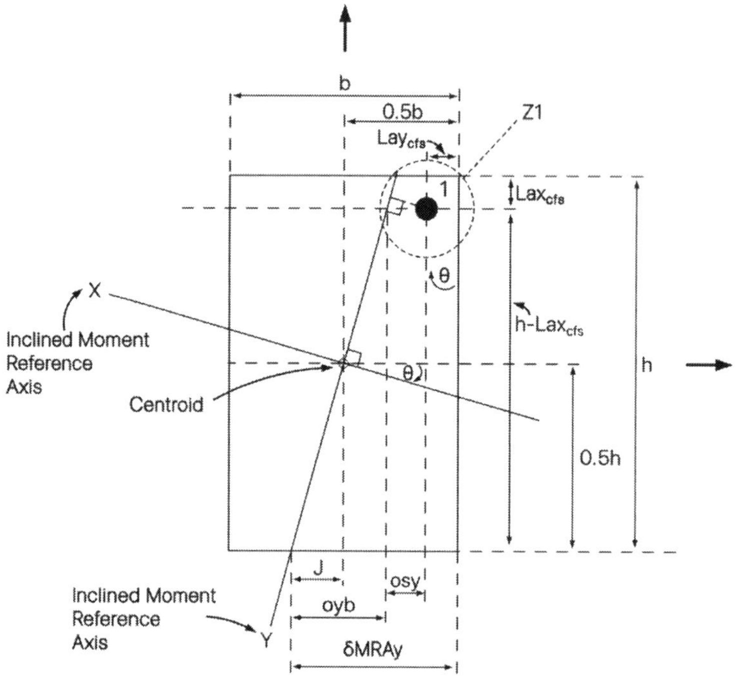

Figure 5.8: The Z1 position for lever arm Lay_{mrsi} for method 2, where the inclined y axis does not surpass bar 1

Derivation of oyb:

$$\tan(\theta) = \frac{\text{opposite}}{\text{adjacent}} = \frac{\text{oyb}}{h - Lax_{cfs}}$$

$$\text{oyb} = \tan(\theta) \cdot (h - Lax_{cfs})$$

Derivation of $\delta MRAy$:

$$\delta MRAy = (0.5 \cdot b) + J$$

Derivation of J:

$$\tan(\theta) = \frac{\text{opposite}}{\text{adjacent}} = \frac{J}{(0.5 \cdot h)}$$

$$J = \tan(\theta) \cdot (0.5 \cdot h)$$

Summary of derived equations

$$Lay_{mrsi} = \cos(\theta) \cdot \text{osy} \tag{5.10}$$

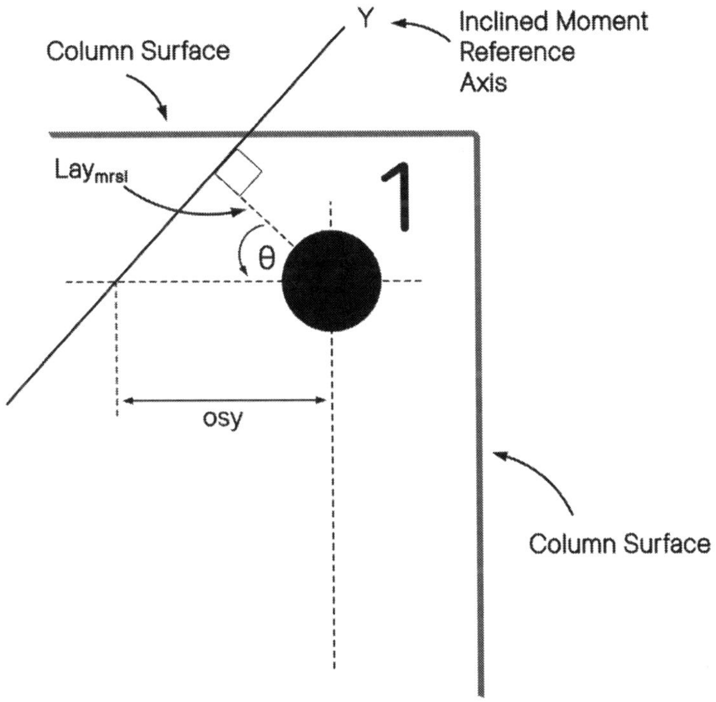

Figure 5.9: Magnification Z1 showing the lever arm Lay_{mrsi} for method 2, where the inclined y axis does not surpass bar 1

$$\text{osy} = \delta MRAy - (\text{oyb} + Lay_{cfs}) \tag{5.11}$$

$$\text{oyb} = \tan(\theta) \cdot (h - Lax_{cfs}) \tag{5.12}$$

$$\delta MRAy = (0.5 \cdot b) + J \tag{5.13}$$

$$J = \tan(\theta) \cdot (0.5 \cdot h) \tag{5.14}$$

Where,

Lay_{mrsi}: Distance from reinforcement centroid to inclined (rotated) moment reference axis y.

Lax_{cfs}: Lever arm for steel for x axis from extreme compression fibre to the centroid of reinforcement bar.

Lay_{cfs}: Lever arm for steel for y axis from extreme compression fibre to the centroid of reinforcement bar.

h: Height of column.

b: Breadth of column.

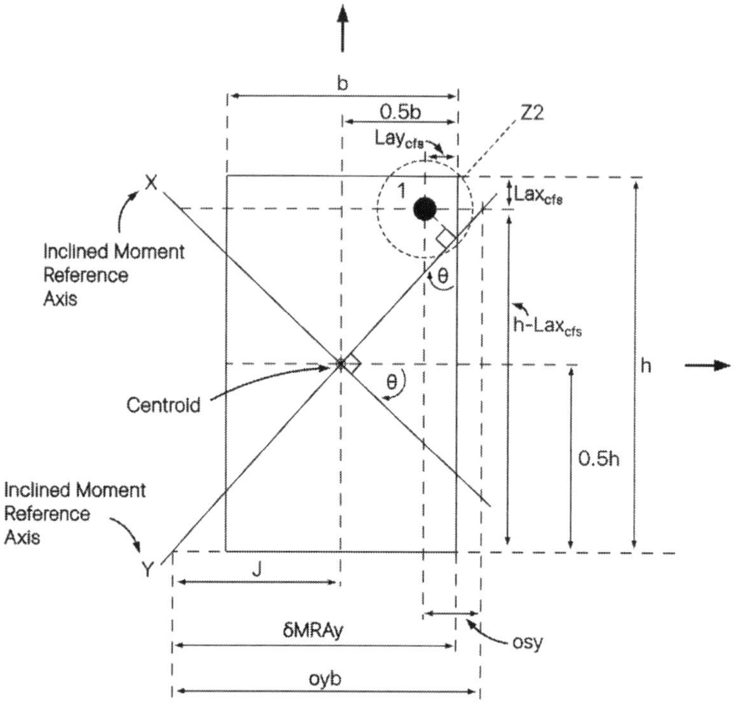

Figure 5.10: The Z2 position for lever arm Lay_{mrsi} for method 2, where the inclined y axis surpasses bar 1

δh: Neutral axis distance along the height.
δb: Neutral axis distance along the breadth.
θ: Inclination angle (rotation angle) of the inclined (rotated) axis relative to the non-inclined (unrotated) axis.

5.3.2.2.2 Example Calculation

Given Data:

$h = 300\,\text{mm}$, $b = 225\,\text{mm}$, $\delta h = 200\,\text{mm}$, $\delta b = 240\,\text{mm}$.

Lever Arm	Bar 1	Bar 2	Bar 3	Bar 4
Lax_{cfs}	40 mm	260 mm	260 mm	40 mm
Lay_{cfs}	40 mm	40 mm	185 mm	185 mm

Calculation of Inclination Angle (θ):

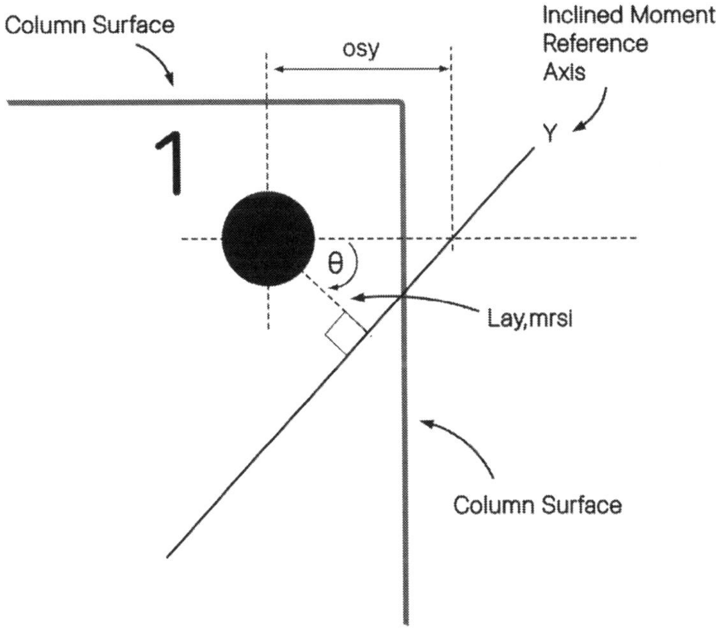

Figure 5.11: Magnification Z2 showing the lever arm Lay_{mrsi} for method 2, where the inclined y axis surpasses bar 1

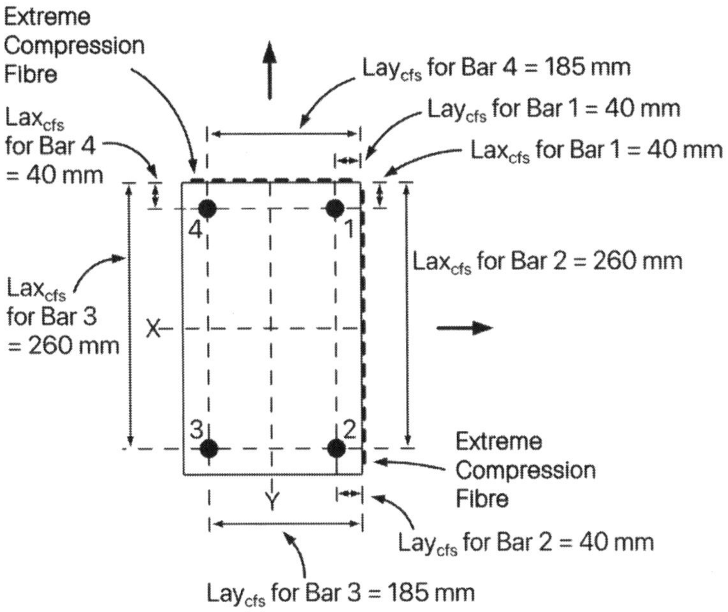

Figure 5.12: The lever arms Lax_{cfs} and Lay_{cfs}

$$\text{Inclination Ratio (IR)} = \frac{\delta h}{\delta b} = \frac{200\,\text{mm}}{240\,\text{mm}} = 0.833333333$$

$$\text{Inclination Angle } (\theta) = \arctan(\text{IR}) = \arctan(0.833333333) = 39.805571092^\circ$$

Note: When using spreadsheet software, ensure that the output of the *arctan* function is converted to degrees by using DEGREES(ATAN(IR)), as some software returns ATAN values in radians by default.

Calculation of J:

$$\begin{aligned} J &= \tan(\theta) \times (0.5 \cdot h) \\ &= \tan(39.805571092^\circ) \times (0.5 \cdot 300\,\text{mm}) \\ &= 0.833333333 \times 150\,\text{mm} \\ &= 125\,\text{mm} \end{aligned}$$

Note: The inclination ratio (IR) is equivalent to $\tan(\theta)$, so it can directly be used in place of $\tan(\theta)$ for simplified calculations.

Calculation of $\delta MRAy$:

$$\begin{aligned} \delta MRAy &= (0.5 \cdot b) + J \\ &= (0.5 \cdot 225\,\text{mm}) + 125\,\text{mm} \\ &= 112.5\,\text{mm} + 125\,\text{mm} \\ &= 237.5\,\text{mm} \end{aligned}$$

Bar 1:

$$\begin{aligned} \text{oyb} &= \tan(\theta) \cdot (h - Lax_{cfs}) \\ &= \tan(39.80557109^\circ) \times (300\,\text{mm} - 40\,\text{mm}) \\ &= 0.833333333 \times 260\,\text{mm} \\ &= 216.667\,\text{mm} \end{aligned}$$

$$\begin{aligned} \text{osy} &= \delta MRAy - (\text{oyb} + Lay_{cfs}) \\ &= 237.5\,\text{mm} - (216.667\,\text{mm} + 40\,\text{mm}) \\ &= -19.167\,\text{mm} \end{aligned}$$

$$\begin{aligned} Lay_{mrsi} &= \cos(\theta) \cdot \text{osy} \\ &= \cos(39.805571092^\circ) \times (-19.167\,\text{mm}) \\ &= -14.724\,\text{mm} \end{aligned}$$

Bar 2:

$$\begin{aligned} \text{oyb} &= \tan(\theta) \cdot (h - Lax_{cfs}) \\ &= 0.833333333 \times (300\,\text{mm} - 260\,\text{mm}) \\ &= 0.833333333 \times 40\,\text{mm} \\ &= 33.333\,\text{mm} \end{aligned}$$

$$\begin{aligned} \text{osy} &= \delta MRAy - (\text{oyb} + Lay_{cfs}) \\ &= 237.500\,\text{mm} - (33.333\,\text{mm} + 40\,\text{mm}) \\ &= 164.167\,\text{mm} \end{aligned}$$

$$\begin{aligned} Lay_{mrsi} &= \cos(\theta) \cdot \text{osy} \\ &= \cos(39.805571092^\circ) \times 164.167\,\text{mm} \\ &= 126.117\,\text{mm} \end{aligned}$$

Bar 3:

$$\begin{aligned} \text{oyb} &= \tan(\theta) \cdot (h - Lax_{cfs}) \\ &= 0.833333333 \times (300\,\text{mm} - 260\,\text{mm}) \\ &= 0.833333333 \times 40\,\text{mm} \\ &= 33.333\,\text{mm} \end{aligned}$$

$$\begin{aligned} \text{osy} &= \delta MRAy - (\text{oyb} + Lay_{cfs}) \\ &= 237.500\,\text{mm} - (33.333\,\text{mm} + 185\,\text{mm}) \\ &= 19.167\,\text{mm} \end{aligned}$$

$$\begin{aligned} Lay_{mrsi} &= \cos(\theta) \cdot \text{osy} \\ &= \cos(39.805571092°) \times 19.167\,\text{mm} \\ &= 14.724\,\text{mm} \end{aligned}$$

Bar 4:

$$\begin{aligned} \text{oyb} &= \tan(\theta) \cdot (h - Lax_{cfs}) \\ &= 0.833333333 \times (300\,\text{mm} - 40\,\text{mm}) \\ &= 0.833333333 \times 260\,\text{mm} \\ &= 216.667\,\text{mm} \end{aligned}$$

$$\begin{aligned} \text{osy} &= \delta MRAy - (\text{oyb} + Lay_{cfs}) \\ &= 237.500\,\text{mm} - (216.667\,\text{mm} + 185\,\text{mm}) \\ &= -164.167\,\text{mm} \end{aligned}$$

$$\begin{aligned} Lay_{mrsi} &= \cos(\theta) \cdot \text{osy} \\ &= \cos(39.805571092°) \times (-164.167\,\text{mm}) \\ &= -126.117\,\text{mm} \end{aligned}$$

5.3.2.3 Summary of Calculations

Distance Type	Bar 1	Bar 2	Bar 3	Bar 4
Lax_{mrsi}	130.918 mm	-38.091 mm	-130.918 mm	38.091 mm
Lay_{mrsi}	-14.724 mm	126.116 mm	14.724 mm	-126.116 mm

5.4 The lever arm La_{nas}

This section focuses on the lever arm La_{nas}, which represents the distance from the neutral axis (NA) to the centroid of the steel reinforcement. By using La_{nas}, the strain at the centroid of the steel reinforcement can be determined based on the strain distribution across the section. The calculation of strain of the reinforcement will be discussed in a later section. This is illustrated in the Figure 5.13.

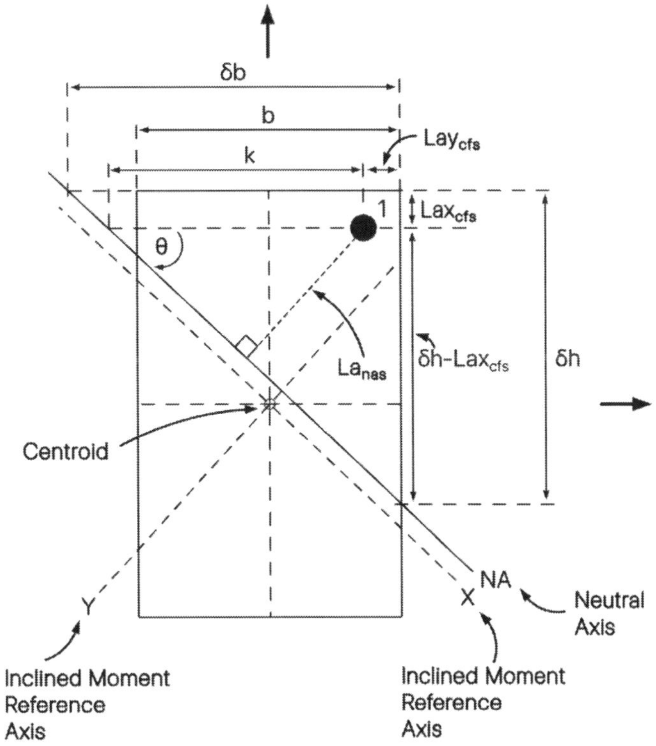

Figure 5.13: The lever arm La_{nas}

5.4.1 Method of Calculation

The lever arm La_{nas}, can be calculated utilizing basic trigonometry. This is shown in Equations 5.15 to 5.19.

$$\sin(\theta) = \frac{\text{opposite}}{\text{hypotenuse}} \tag{5.15}$$

$$\sin(\theta) = \frac{La_{nas}}{k} \tag{5.16}$$

$$La_{nas} = \sin(\theta) \cdot k \tag{5.17}$$

$$k + Lay_{cfs} = \left(\frac{\delta b}{\delta h}\right) - (\delta h - Lax_{cfs}) \tag{5.18}$$

$$k = \left(\frac{\delta b}{\delta h} \cdot (\delta h - Lax_{cfs}) \right) - Lay_{cfs} \tag{5.19}$$

Where,

θ: Inclination angle of the neutral axis.

La_{nas}: Distance from the neutral axis (NA) to the centroid of reinforcement.

Lax_{cfs}: Lever arm for steel for x axis from extreme compression fibre to the centroid of reinforcement bar.

Lay_{cfs}: Lever arm for steel for y axis from extreme compression fibre to the centroid of reinforcement bar.

δh: Neutral axis distance along the height.

δb: Neutral axis distance along the breadth.

5.4.2 Example Calculation

Given Data:

$h = 300\,\text{mm}$, $b = 225\,\text{mm}$, $\delta h = 200\,\text{mm}$, $\delta b = 240\,\text{mm}$, $\theta = 39.805571092°$

Lever Arm	Bar 1	Bar 2	Bar 3	Bar 4
Lax_{cfs}	40 mm	260 mm	260 mm	40 mm
Lay_{cfs}	40 mm	40 mm	185 mm	185 mm

Bar 1:

$$Lax_{cfs} = 40\,\text{mm}$$
$$Lay_{cfs} = 40\,\text{mm}$$

Calculation of k:

$$\begin{aligned} k &= \left(\frac{\delta b}{\delta h} \right) \cdot (\delta h - Lax_{cfs}) - Lay_{cfs} \\ &= \left(\frac{240\,\text{mm}}{200\,\text{mm}} \right) \cdot (200\,\text{mm} - 40\,\text{mm}) - 40\,\text{mm} \\ &= 152.000\,\text{mm} \end{aligned}$$

Calculation of La_{nas}:

$$
\begin{aligned}
La_{nas} &= \sin(\theta) \cdot k \\
&= \sin(39.805571092^\circ) \times 152.000\,\text{mm} \\
&= 97.308\,\text{mm}
\end{aligned}
$$

Bar 2:

$$
\begin{aligned}
Lax_{cfs} &= 260\,\text{mm} \\
Lay_{cfs} &= 40\,\text{mm}
\end{aligned}
$$

Calculation of k:

$$
\begin{aligned}
k &= \left(\frac{\delta b}{\delta h}\right) \cdot (\delta h - Lax_{cfs}) - Lay_{cfs} \\
&= \left(\frac{240\,\text{mm}}{200\,\text{mm}}\right) \cdot (200\,\text{mm} - 260\,\text{mm}) - 40\,\text{mm} \\
&= -112.000\,\text{mm}
\end{aligned}
$$

Calculation of La_{nas}:

$$
\begin{aligned}
La_{nas} &= \sin(\theta) \cdot k \\
&= \sin(39.805571092^\circ) \times (-112\,\text{mm}) \\
&= -71.701\,\text{mm}
\end{aligned}
$$

Bar 3:

$$
\begin{aligned}
Lax_{cfs} &= 260\,\text{mm} \\
Lay_{cfs} &= 185\,\text{mm}
\end{aligned}
$$

Calculation of k:

$$
\begin{aligned}
k &= \left(\frac{\delta b}{\delta h}\right) \cdot (\delta h - Lax_{cfs}) - Lay_{cfs} \\
&= \left(\frac{240\,\text{mm}}{200\,\text{mm}}\right) \times (200\,\text{mm} - 260\,\text{mm}) - 185\,\text{mm} \\
&= -257.000\,\text{mm}
\end{aligned}
$$

Calculation of La_{nas}:

$$
La_{nas} = \sin(\theta) \cdot k
$$

$$= \sin(39.805571092^\circ) \times (-257\,\text{mm})$$
$$= -164.527\,\text{mm}$$

Bar 4:

$$Lax_{cfs} = 40\,\text{mm}$$
$$Lay_{cfs} = 185\,\text{mm}$$

Calculation of k:

$$k = \left(\frac{\delta b}{\delta h}\right) \cdot (\delta h - Lax_{cfs}) - Lay_{cfs}$$
$$= \left(\frac{240\,\text{mm}}{200\,\text{mm}}\right) \times (200\,\text{mm} - 40\,\text{mm}) - 185\,\text{mm}$$
$$= 7.000\,\text{mm}$$

Calculation of La_{nas}:

$$La_{nas} = \sin(\theta) \cdot k$$
$$= \sin(39.805571092^\circ) \times 7\,\text{mm}$$
$$= 4.481\,\text{mm}$$

Summary of Calculations

Lever Arm Type	**Bar 1**	**Bar 2**	**Bar 3**	**Bar 4**
La_{nas}	97.308 mm	-71.701 mm	-164.527 mm	4.481 mm

Chapter 6:

Resistive Axial Force and Moments in Steel

This chapter describes the resistive axial force of steel (N_s), and resistive moments of steel (M_{xsn}, M_{ysn}, M_{xsi}, and M_{ysi}), within a rectangular column subjected to biaxial bending. The resistive axial force (N_s) represents the force in the steel reinforcement resisting compression or tension along the inclined or non-inclined x and y moment reference axes of the column. Moreover, the resistive moments of steel (M_{xsn}, M_{ysn}, M_{xsi}, and M_{ysi}) represent the contributions of steel reinforcement in resisting bending moments about the inclined or non-inclined x and y moment reference axes, respectively.

6.1 Resistive Axial Force in Steel (N_s)

In a reinforced concrete column under biaxial bending, steel reinforcement plays a crucial role in resisting axial loads. While concrete provides compressive resistance, steel reinforcement provides both compressive and tensile resistance.

Moreover, when determining concrete forces and moments in an earlier section, the reinforcement area was not subtracted from the area of concrete compressive zone. Thereby, to ensure accuracy, the concrete stress at the compressive steel locations is subtracted to avoid overestimating the contribution of concrete. This reduction is applied in the calculation of the steel stress (σyd_{bc}) for each bar in consideration. This ensures a balanced and accurate determination of forces.

6.1.1 Strain in Steel

The strain in steel reinforcement plays a critical role in determining its contribution to the overall resistive behaviour of a reinforced concrete column under biaxial bending. Steel strain defines the extent of deformation within the reinforcement, directly influencing the resistive forces and moments generated by the steel.

The steel strain can be calculated using two main methods, non-inclined (unrotated) axis method and the inclined (rotated) axis method. Both methods yield the exact same results, and the choice between them depends on preference. However, the inclined method aligns more naturally with the biaxial geometry of the column. Thereby, the inclined method is considered the more intuitive approach.

6.1.1.1 Method 1 - Non-inclined (Un-rotated) Axes Method

In this method the strain is calculated by utilizing the non-inclined moment reference axes and neutral axis.

6.1.1.1.1 Method of Calculation

Before using Equations 6.1 and Equation 6.2, it is important to determine whether the bar is in compression or tension. This can be established by evaluating the sign of the strain of the bar before the strain limit check is applied (εbcu). If the strain of the bar in consideration (εbcu) is negative (−), the bar is in tension. Conversely, if the strain of the bar in consideration (εbcu) is positive (+), the bar can be considered to be in compression. This is illustrated in Figure 6.1.

$$\varepsilon yd_{bc} = \min\left[(\varepsilon_{bcu})\,;\,(+\varepsilon_{ms})\right] \quad \text{If the bar is in compression (+)} \tag{6.1}$$

$$\varepsilon yd_{bc} = \max\left[(\varepsilon_{bcu})\,;\,(-\varepsilon_{ms})\right] \quad \text{If the bar is in tension (−)} \tag{6.2}$$

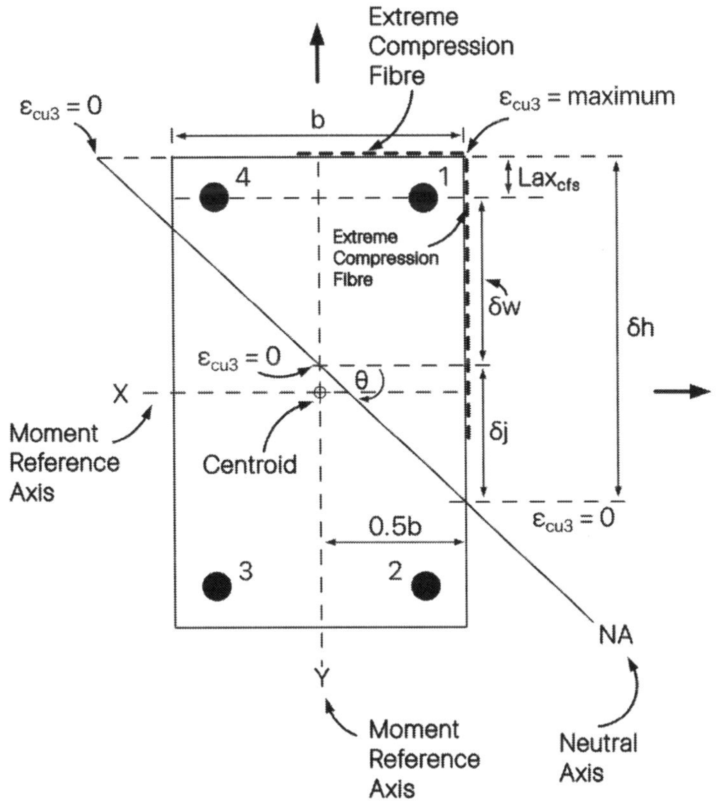

Figure 6.1: Strain distribution relative to neutral axis (NA)

$$\varepsilon_{ms} = \frac{f_{yd}}{E} \tag{6.3}$$

$$f_{yd} = \frac{f_{yk}}{\gamma_s} \tag{6.4}$$

$$\varepsilon_{bcu} = \varepsilon_{conx} + \varepsilon_{cony} \tag{6.5}$$

$$\varepsilon_{conx} = \varepsilon_{gx} \cdot \delta w \tag{6.6}$$

$$\varepsilon_{cony} = \varepsilon_{gy} \cdot Lay_{mrs} \tag{6.7}$$

$$\varepsilon_{gx} = \frac{\varepsilon_{cu3}}{\delta h} \tag{6.8}$$

$$\varepsilon_{gy} = \frac{\varepsilon_{cu3}}{\delta b} \tag{6.9}$$

$$\delta w = \delta h - (Lax_{cfs} + \delta j) \tag{6.10}$$

$$\delta j = \tan(\theta) \cdot (0.5 \cdot b) \tag{6.11}$$

Where,

εyd_{bc}: Design strain of the bar in consideration.

ε_{bcu}: Strain of the bar in consideration before the strain limit check is applied.

ε_{cu3}: Maximum design strain of concrete (typically 0.0035. This applies to concrete with f_{ck} up to 50 N/mm^2).

ε_{ms}: Maximum yield strain of steel (typically 0.00217392).

E: Young's modulus of steel (e.g.: typically around 200×10^3 MPa or 200×10^3 N/mm^2).

f_{yk}: Characteristic yield strength of steel (typically 500 MPa or 500 N/mm^2).

f_{yd}: Design yield strength of steel.

γs: Partial safety factor for steel (typically 1.15).

ε_{conx} = Strain contribution along moment reference axis x.

ε_{cony} = Strain contribution along moment reference axis y.

ε_{gx} = Concrete strain gradient from extreme compression fibre to neutral axis, along x axis (concrete strain per unit distance).

ε_{gy} = Concrete strain gradient from extreme compression fibre to neutral axis, along y axis (concrete strain per unit distance).

δb: Neutral axis distance along the breadth.

δh: Neutral axis distance along the height.

6.1.1.1.2 Example Calculations

Scenario 1: Column is not in pure compression

Given Data:

$h = 300\,\text{mm}$, $b = 225\,\text{mm}$, $\delta h = 200\,\text{mm}$, $\delta b = 240\,\text{mm}$, $\varepsilon_{cu3} = 0.0035$, $f_{yk} = 500\,\text{N/mm}^2$, $\gamma s = 1.15$, $E = 200000\,\text{N/mm}^2$.

Lever Arm Type	Bar 1	Bar 2	Bar 3	Bar 4
Lax_{cfs}	40 mm	260 mm	260 mm	40 mm
Lay_{mrs}	72.5 mm	72.5 mm	-72.5 mm	-72.5 mm

Calculation of f_{yd}:

$$\begin{aligned} f_{yd} &= \frac{f_{yk}}{\gamma_s} \\ &= \frac{500\,\mathrm{N/mm^2}}{1.15} \\ &= 434.783\,\mathrm{N/mm^2} \end{aligned}$$

Calculation of ε_{ms}:

$$\begin{aligned} \varepsilon_{ms} &= \frac{f_{yd}}{E} \\ &= \frac{434.783\,\mathrm{N/mm^2}}{200000} \\ &= 0.002173920 \end{aligned}$$

Calculation of IR:

$$\begin{aligned} IR &= \frac{\delta h}{\delta b} \\ &= \frac{200\,\mathrm{mm}}{240\,\mathrm{mm}} \\ &= 0.833333333 \end{aligned}$$

Angle θ:

$$\begin{aligned} \theta &= \arctan(IR) \\ &= \arctan(0.833333333) \\ &= 39.805571092^\circ \end{aligned}$$

Calculation of δj:

$$\begin{aligned} \delta j &= \tan(\theta) \cdot (0.5 \cdot b) \\ &= \tan(39.805571092^\circ) \times (0.5 \times 225\,\mathrm{mm}) \\ &= 0.833333333 \times 112.5\,\mathrm{mm} \\ &= 93.750\,\mathrm{mm} \end{aligned}$$

Calculation of Concrete Strain Gradients Along x and y:

$$\varepsilon_{gx} = \frac{\varepsilon_{cu3}}{\delta h}$$
$$= \frac{0.0035}{200\,\text{mm}}$$
$$= 0.000017500$$

$$\varepsilon_{gy} = \frac{\varepsilon_{cu3}}{\delta b}$$
$$= \frac{0.0035}{240\,\text{mm}}$$
$$= 0.000014583$$

Calculation of δ_w for Each Bar:

$$\delta w = \delta h - (Lax_{cfs} + \delta j)$$

Bar 1:

$$\delta w = 200\,\text{mm} - (40\,\text{mm} + 93.750\,\text{mm})$$
$$= 66.250\,\text{mm}$$

Bar 2:

$$\delta w = 200\,\text{mm} - (260\,\text{mm} + 93.750\,\text{mm})$$
$$= -153.750\,\text{mm}$$

Bar 3:

$$\delta w = 200\,\text{mm} - (260\,\text{mm} + 93.750\,\text{mm})$$
$$= -153.750\,\text{mm}$$

Bar 4:

$$\delta w = 200\,\text{mm} - (40\,\text{mm} + 93.750\,\text{mm})$$
$$= 66.250\,\text{mm}$$

Calculation of ε_{conx} and ε_{cony} for Each Bar:

$$\varepsilon_{conx} = \varepsilon_{gx} \cdot \delta w$$

$$\varepsilon_{cony} = \varepsilon_{gy} \cdot Lay_{mrs}$$

Bar 1:

$$\varepsilon_{conx} = \left(\frac{0.0035}{200\,\text{mm}}\right) \times 66.250\,\text{mm}$$

$$= +0.001159375$$

$$\varepsilon_{cony} = \left(\frac{0.0035}{240\,\text{mm}}\right) \times 72.500\,\text{mm}$$

$$= +0.001057292$$

Bar 2:

$$\varepsilon_{conx} = \left(\frac{0.0035}{200\,\text{mm}}\right) \times -153.750\,\text{mm}$$

$$= -0.002690625$$

$$\varepsilon_{cony} = \left(\frac{0.0035}{240\,\text{mm}}\right) \times 72.500\,\text{mm}$$

$$= +0.001057292$$

Bar 3:

$$\varepsilon_{conx} = \left(\frac{0.0035}{200\,\text{mm}}\right) \times -153.750\,\text{mm}$$

$$= -0.002690625$$

$$\varepsilon_{cony} = \left(\frac{0.0035}{240\,\text{mm}}\right) \times -72.500\,\text{mm}$$

$$= -0.001057292$$

Bar 4:

$$\varepsilon_{conx} = \left(\frac{0.0035}{200\,\text{mm}}\right) \times 66.250\,\text{mm}$$

$$= +0.001159375$$

$$\varepsilon_{cony} = \left(\frac{0.0035}{240\,\text{mm}}\right) \times -72.500\,\text{mm}$$
$$= -0.001057292$$

Calculation of ε_{bcu}:

$$\varepsilon_{bcu} = \varepsilon_{conx} + \varepsilon_{cony}$$

Bar 1:

$$\begin{aligned}\varepsilon_{bcu} &= (+0.001159375) + (+0.001057292)\\ &= +0.002216667\end{aligned}$$

Bar 2:

$$\begin{aligned}\varepsilon_{bcu} &= (-0.002690625) + (+0.001057292)\\ &= -0.001633333\end{aligned}$$

Bar 3:

$$\begin{aligned}\varepsilon_{bcu} &= (-0.002690625) + (-0.001057292)\\ &= -0.003747917\end{aligned}$$

Bar 4:

$$\begin{aligned}\varepsilon_{bcu} &= (+0.001159375) + (-0.001057292)\\ &= +0.000102083\end{aligned}$$

Calculation of $\varepsilon_{yd,bc}$:
For the bar 1 and 4, the ε_{bcu} is positive (+), thereby those bars are in compression.
For the bar 2 and 3, the ε_{bcu} is negative (−), thereby those bars are in tension.

$$\begin{aligned}\varepsilon yd_{bc} &= \min(\varepsilon_{bcu}, +\varepsilon_{ms}) \quad \text{(for compressive bars)}\\ \varepsilon yd_{bc} &= \max(\varepsilon_{bcu}, -\varepsilon_{ms}) \quad \text{(for tensile bars)}\end{aligned}$$

Bar 1 (Compression):

$$\begin{aligned}\varepsilon yd_{bc} &= \min(0.002216667, 0.002173920)\\ &= +0.002173920\end{aligned}$$

Bar 2 (Tension):

$$\begin{aligned}\varepsilon yd_{bc} &= \max(-0.001633333, -0.002173920)\\ &= -0.001633333\end{aligned}$$

Bar 3 (Tension):

$$\begin{aligned}\varepsilon yd_{bc} &= \max(-0.003747917, -0.002173920)\\ &= -0.002173920\end{aligned}$$

Bar 4 (Compression):

$$\begin{aligned}\varepsilon yd_{bc} &= \min(0.000102083, 0.002173920)\\ &= +0.000102083\end{aligned}$$

Summary of Calculations:

Strain	Bar 1	Bar 2	Bar 3	Bar 4
εyd_{bc}	+0.002173920	-0.001633333	-0.002173920	+0.000102083

6.1.1.2 Method 2 - Inclined (Rotated) Axis Method

In this method, the strain is calculated by utilizing the inclined moment reference axis.

6.1.1.2.1 Method of Calculation

Based on the principle of strain compatibility, which states that steel and concrete, being bonded together, deform simultaneously under applied loads, the surrounding concrete and the reinforcement bar elongate or contract in unison. Therefore, the concrete strain at the bar location is taken as the strain in the reinforcement bar. This ensures that the strain at the bar location reflects this compatibility.

However, this design assumption is only valid as long as the concrete remains uncracked under tensile loads. Concrete has a very low tensile strength compared to its compressive strength, allowing it to sustain only a small amount of tensile strain before cracking. For example, in 30 N/mm^2 concrete, the tensile strength (f_{ctm}) is approximately 3.2 N/mm^2 (based on standard relations for concrete strength classes), and with an elastic modulus (E_{cm}) of around 31,000 N/mm^2 (Eurocode 2, Table 3.1), the tensile strain at failure can be calculated as 0.00010323. This small strain limit means that concrete cannot bear significant tensile strain and is often considered to have almost no tension-bearing capacity. As soon as the tensile strain in the concrete exceeds this threshold, it cracks, as it cannot elongate further to resist the applied tension. At this point, the steel reinforcement, with its much higher tensile strain capacity takes over the tensile load.

Due to this phenomenon, in real-life reinforced concrete columns, the tension zones inevitably develop physical microscopic cracks in concrete, caused by the strain in the steel. These microscopic cracks or gaps open and close at a microscopic level as the loads on the column vary, and they may not fully close once they have opened. These cracks remain microscopic and are typically not visible to the naked eye unless the concrete is heavily stressed or poorly designed. However, strain compatibility remains a useful design assumption in uncracked regions and for simplified models.

For the Equations 6.12 and 6.13, to determine whether the bar in consideration is in compression or tension, examine the sign of distance from the neutral axis (NA) to the centroid of reinforcement (La_{nas}). If La_{nas} is negative (-), the bar is in tension. Conversely, if La_{nas} is positive (+), the bar can be considered to be in compression. If the bar in consideration is in tension, the maximum value (close to zero in negative range) between two values should be selected. For example, if the values are '-0.01' and '-0.0001', then the maximum strain value (closer to zero in negative range) '-0.0001', should be chosen. Conversely, if the bar is in compression, the least value (closer to zero in positive range) should be chosen. For example, if the values are '+0.01' and '+0.0001', then '+0.0001' should be chosen. This ensures that the design stress reflects the true stress state of the bar.

$$\varepsilon yd_{bc} = \min\left[\left(\frac{\varepsilon_{cu3}}{La_{nac}} \cdot La_{nas}\right); (+\varepsilon_{ms})\right] \quad \text{If the bar is in compression (+)} \tag{6.12}$$

$$\varepsilon yd_{bc} = \max\left[\left(\frac{\varepsilon_{cu3}}{La_{nac}} \cdot La_{nas}\right); (-\varepsilon_{ms})\right] \quad \text{If the bar is in tension (-)} \tag{6.13}$$

$$\varepsilon_{ms} = \frac{f_{yd}}{E_s} \tag{6.14}$$

$$f_{yd} = \frac{f_{yk}}{\gamma_s} \tag{6.15}$$

Where,

$\varepsilon_{yd,bc}$: Design strain of the bar in consideration.

La_{nac}: Distance from the neutral axis (NA) to the extreme compression fibre intersection point (ECFIP).

La_{nas}: Distance from the neutral axis (NA) to the centroid of reinforcement.

ε_{cu3}: Maximum design strain of concrete (typically 0.0035. This applies to concrete with f_{ck} up to 50 N/mm^2).

ε_{ms}: Maximum design yield strain of steel (typically 0.00217392).

E_s: Young's modulus of steel (e.g., typically around 200×10^3 MPa).

f_{yk}: Characteristic yield strength of steel (typically 500 MPa or 500 N/mm^2).

f_{yd}: Design yield strength of steel.

γ_s: Partial safety factor for steel (typically 1.15).

6.1.1.2.2 Example Calculations

Scenario 1: Column is not in pure compression

Given Data:

Lever Arm Type	Bar 1	Bar 2	Bar 3	Bar 4
La_{nas}	97.308 mm	-71.701 mm	-164.527 mm	4.481 mm
La_{nac}	153.644 mm	153.644 mm	153.644 mm	153.644 mm

Calculation of f_{yd}:

$$\begin{aligned} f_{yd} &= \frac{f_{yk}}{\gamma_s} \\ &= \frac{500}{1.15} \\ &= 434.783\,\mathrm{N/mm^2} \end{aligned}$$

Calculation of ε_{ms}:

$$\begin{aligned} \varepsilon_{ms} &= \frac{f_{yd}}{E} \\ &= \frac{434.783\,\mathrm{N/mm^2}}{200000\,\mathrm{N/mm^2}} \\ &= 0.00217392 \end{aligned}$$

Calculation of εyd_{bc} for Each Bar:

For the bar 1 and 4 the La_{nas} is positive (+), thereby those bars are in compression.

For the bar 2 and 3 the La_{nas} is negative (−), thereby those bars are in tension.

For bars in compression (+):

$$\varepsilon yd_{bc} = \min\left[\left(\frac{\varepsilon_{cu3}}{La_{nac}} \cdot La_{nas}\right); (+\varepsilon_{ms})\right]$$

For bars in tension (−):

$$\varepsilon yd_{bc} = \max\left[\left(\frac{\varepsilon_{cu3}}{La_{nac}}\cdot -La_{nas}\right);\, (-\varepsilon_{ms})\right]$$

Bar 1 (Compression):

$$\varepsilon yd_{bc} = \min\left[\left(\frac{0.0035}{153.644\,\text{mm}} \times 97.308\,\text{mm}\right);\, (+0.00217392)\right]$$

$$= \min\,[+0.002216667;\, +0.00217392]$$

$$= +0.002173920$$

Bar 2 (Tension):

$$\varepsilon yd_{bc} = \max\left[\left(\frac{0.0035}{153.644\,\text{mm}} \times -71.701\,\text{mm}\right);\, (-0.00217392)\right]$$

$$= \max\,[-0.001633333;\, -0.00217392]$$

$$= -0.001633333$$

Bar 3 (Tension):

$$\varepsilon yd_{bc} = \max\left[\left(\frac{0.0035}{153.644\,\text{mm}} \times -164.527\,\text{mm}\right);\, (-0.00217392)\right]$$

$$= \max\,[-0.003747917;\, -0.00217392]$$

$$= -0.002173920$$

Bar 4 (Compression):

$$\varepsilon yd_{bc} = \min\left[\left(\frac{0.0035}{153.644\,\text{mm}} \times 4.481\,\text{mm}\right);\, (+0.00217392)\right]$$

$$= \min\,[+0.000102083;\, +0.00217392]$$

$$= +0.000102083$$

Summary of the Results:

Strain Type	Bar 1	Bar 2	Bar 3	Bar 4
εyd_{bc}	+0.002173920	-0.001633333	-0.002173920	+0.000102083

Note 1: An important observation is that both Method 1 (Non-inclined Axis) and Method 2 (Inclined Axis) lead to the same strain values for each reinforcement bar. While the approaches differ in reference axis orientation, the final strain values must be identical confirming the consistency of the methodology.

Note 2: Although this example considers only four bars for simplicity, the same process applies to any number of reinforcement bars. Regardless of the number of bars, the method remains valid for determining bar stresses and strains.

6.1.1.3 If the Column is in Pure Compression

Columns experience behaviour cases where the entire cross-section is completely in compression, also known as pure compression. In such unique scenarios, Eurocode 2 (EN 1992-1-1), imposes a maximum strain limit.

6.1.1.3.1 Method of Calculation

Specifically, Eurocode limits the concrete compressive strain to εc_3 (0.00175 for Concrete strength grade $\leq$ C50/60 N/mm2) in approximately concentric compression (i.e., when eccentricity ($e_d \leq$ 0.1h). This is mentioned in the Eurocode 2 (EN 1992-1-1), Clause 6.1 (5). That means this applies when the column is subjected to small eccentricity. Thereby, this typically applies to concrete in compression.

However, strain was not explicitly calculated to determine the compressive stress of concrete in previous sections, instead, the design compressive strength of concrete (f_{cd}) was utilized due to the adoption of a simplified bilinear stress-strain model throughout this book.

Nevertheless, the strain limitation remains inherently active. Consequently, to maintain strain compatibility, the same restriction must be applied to the reinforcement steel, since both materials share the same strain field in compression. Consequently, even though the clause explicitly governs concrete behaviour, the steel reinforcement is effectively subjected to the same strain limit.

In the pure compression range, a simplified, strain-compatible yet conservative approach was utilized for the concrete strain limit which limits the strain at 0.00175 in accordance with Eurocode 2. This avoids additional complexity and improves the overall curvature of the interaction diagram.

Equation 6.17 includes the derived expression and the strain cap of 0.00175. This strain value is specified in Eurocode 2 (EN 1992-1-1), Table 3.1. Moreover, since we adopted the simplified approach, the column is considered to be in pure compression if the Equation 6.16, is satisfied.

$$A_{cc} = (h \cdot b) \tag{6.16}$$

Where,
Acc: Area of the compression zone of concrete.
b: Breadth of the column cross-section.
h: Height of the column cross-section.

$$\varepsilon yd_{bcp} = \min\left[(\varepsilon yd_{bc})\,;\,(\varepsilon_{c3})\right] \tag{6.17}$$

Where,
εyd_{bcp}: Design strain of the bar when the column is subjected to pure compression.
εyd_{bc}: Design strain of the bar in consideration.

ε_{c3}: Compressive strain at the peak stress in the bi-linear stress-strain diagram (typically 0.00175). This applies to concrete with f_{ck} up to 50 N/mm^2).

Note: When the column is in pure compression, the strain εyd_{bcp} should be utilized for design stress (σyd_{bc}) calculations of the reinforcement instead of εyd_{bc}.

6.1.1.3.2 Example Calculation

Given Data: $\varepsilon yd_{bc} = 0.00180$, $\varepsilon_{c3} = 0.00175$.

Calculation:

$$\begin{aligned}\varepsilon yd_{bcp} &= \min\left[\varepsilon yd_{bc}; \varepsilon_{c3}\right]\\ &= \min\left[0.00180; 0.00175\right]\\ &= 0.00175\end{aligned}$$

6.1.2 Design Stress in Steel (σyd_{bc})(Maximum Allowable Stress in Steel)

The design stress for each steel bar (σyd_{bc}) in the reinforcement is calculated based on the strain in the steel and material properties, ensuring that the steel can resist the applied loads without yielding.

6.1.2.1 Method of Calculation

If the bar in consideration is in tension, the least value between the product of Young's modulus (E) and the design yield strain (σyd_{bc}) and the negative design yield strength ($-f_{yd}$) should be selected. For example, if the values are '-434.784' and '-5', then the maximum stress value (closer to zero in negative range) '−5', should be chosen. Conversely, if the bar is in compression, the least value (close to zero in positive range) between the product of Young's modulus (E) and the design yield strain (σyd_{bc}) and the positive design yield strength ($+f_{yd}$) should be selected. For example, if the values are '+434.784' and '+5', then '+5' should be chosen. This ensures that the design stress reflects the true stress state of the bar.

The bar can be considered in tension or compression depending on whether the distance from the neutral axis (NA) to the centroid of the reinforcement (La_{nas}) is negative (−) or positive (+). If it is negative (−), the bar is in tension, and if it is positive (+), the bar is in compression.

Moreover, the concrete stress at the bar location is reduced because the concrete area at the bar location, which should have been excluded from the total concrete area of the section, was not subtracted earlier. This omission was made to avoid introducing slight errors when determining the centroid of the compressive concrete area in a previous section. To account for this, the adjustment is now made by reducing the concrete stress (σcd) from the compressive bar stress (σyd_{bc}), ensuring accuracy. This reduction is applied only to the compressive bar stress, not to the tensile bar stress. This is shown in the example calculation. Moreover, the calculation formulas are shown in the Equations 6.18 and 6.21. The Sigma (σ), Epsilon (ε), and Gamma (γ) are Greek letters utilized for notation.

$$\sigma yd_{bc} = \max((E_s \cdot -\varepsilon yd_{bc}); (-f_{yd})) \quad \text{If the bar is in tension} \tag{6.18}$$

$$\sigma yd_{bc} = \min((E_s \cdot +\varepsilon yd_{bc}); (+f_{yd} - \sigma cd)) \quad \text{If the bar is in compression} \tag{6.19}$$

$$f_{yd} = \frac{f_{yk}}{\gamma_s} \tag{6.20}$$

$$\sigma cd = \frac{f_{ck}}{\gamma_c} \tag{6.21}$$

Where,

σyd_{bc}: Design stress in each steel bar in consideration (maximum allowable stress in steel).
E_s: Young's modulus of steel (e.g., typically around $200 \times 10^3\,\mathrm{N/mm^2}$).
εyd_{bc}: Design yield strain of steel for each bar in consideration.
f_{yk}: Characteristic yield strength of steel (typically 500 MPa or $500\,\mathrm{N/mm^2}$).
f_{yd}: Design yield strength of steel.
γ_s: Partial safety factor for steel (typically 1.15).
γ_c: Partial safety factor for concrete (typically 1.5).
σcd: Design stress of concrete (i.e., f_{cd}).

6.1.2.2 Example Calculation

Given Data:

$f_{ck} = 30\,\mathrm{N/mm^2}, \quad E_s = 200000\,\mathrm{N/mm^2}, \quad \gamma_s = 1.15, \quad \gamma_c = 1.5.$

Strain Type	Bar 1	Bar 2	Bar 3	Bar 4
εyd_{bc}	+0.002173920	-0.001633333	-0.002173920	+0.000102083

Calculation:

If the bar is in tension: $\sigma yd_{bc} = \max\left[(E_s \cdot (-\varepsilon yd_{bc}))\,;(-f_{yd})\right]$
If the bar is in compression: $\sigma yd_{bc} = \min\left[(E_s \cdot (+\varepsilon yd_{bc}))\,;(+f_{yd})\right] - \sigma_{cd}$

The equation and the calculation for design strength of steel reinforcement (fyd), is stated in the Clause 3.2.7 (2) in Eurocode 2 (EN 1992-1-1) as follows.

$$
\begin{aligned}
f_{yd} &= \frac{f_{yk}}{\gamma_s} \\
&= \frac{500}{1.15} \\
&= 434.782\,\mathrm{N/mm^2}
\end{aligned}
$$

The equation for design compressive strength of concrete (f_{cd}), which is utilized as design stress (σ_{cd}) of concrete in this calculation, is stated in the Clause 3.1.6 (1) in Eurocode 2 (EN 1992-1-1) as follows.

$$
\begin{aligned}
f_{cd} &= \frac{\alpha_{cc} \cdot f_{ck}}{\gamma_c} \\
&= \frac{0.85 \cdot 30\,\mathrm{N/mm^2}}{1.5} \\
&= 17\,\mathrm{N/mm^2}
\end{aligned}
$$

Bar 1 (Compression):

$$
\begin{aligned}
\sigma yd_{bc} &= \min\left[(E_s \cdot (+\varepsilon yd_{bc}))\,;(+f_{yd})\right] - \sigma_{cd} \\
&= \min\left[(200000 \cdot 0.002173920)\,;\left(434.783\,\mathrm{N/mm^2}\right)\right] - 17\,\mathrm{N/mm^2} \\
&= \min\left[434.784\,\mathrm{N/mm^2};\left(434.783\,\mathrm{N/mm^2}\right)\right] - 17\,\mathrm{N/mm^2} \\
&= 434.783\,\mathrm{N/mm^2} - 17\,\mathrm{N/mm^2} \\
&= 417.783\,\mathrm{N/mm^2}
\end{aligned}
$$

Bar 2 (Tension):

$$
\begin{aligned}
\sigma yd_{bc} &= \max\left[(E_s \cdot (-\varepsilon yd_{bc}))\,;(-f_{yd})\right] \\
&= \max\left[(200000 \cdot -0.001633333)\,;\left(-434.783\,\mathrm{N/mm^2}\right)\right] \\
&= \max\left[-326.667\,\mathrm{N/mm^2};\left(-434.783\,\mathrm{N/mm^2}\right)\right] \\
&= -326.667\,\mathrm{N/mm^2}
\end{aligned}
$$

Bar 3 (Tension):

$$
\sigma yd_{bc} = \max\left[(E_s \cdot (-\varepsilon yd_{bc}))\,;(-f_{yd})\right]
$$

$$= \max\left[(200000 \cdot -0.002173920)\,;\left(-434.783\,\mathrm{N/mm^2}\right)\right]$$

$$= \max\left[-434.784\,\mathrm{N/mm^2};\left(-434.783\,\mathrm{N/mm^2}\right)\right]$$

$$= -434.783\,\mathrm{N/mm^2}$$

Bar 4 (Compression):

$$\sigma yd_{bc} = \min\left[(E_s \cdot (+\varepsilon yd_{bc}))\,;(+f_{yd})\right] - \sigma_{cd}$$

$$= \min\left[(200000 \cdot 0.000102083)\,;\left(434.783\,\mathrm{N/mm^2}\right)\right] - 17\,\mathrm{N/mm^2}$$

$$= \min\left[20.417\,\mathrm{N/mm^2};\left(434.783\,\mathrm{N/mm^2}\right)\right] - 17\,\mathrm{N/mm^2}$$

$$= 20.417\,\mathrm{N/mm^2} - 17\,\mathrm{N/mm^2}$$

$$= 3.417\,\mathrm{N/mm^2}$$

Summary of Calculations

Stress	Bar 1	Bar 2	Bar 3	Bar 4
σyd_{bc}	417.783 N/mm^2	-326.667 N/mm^2	-434.783 N/mm^2	3.417 N/mm^2

6.1.3 Area of Steel (A_s)

The area of steel (As) for this calculation depends on the selected total steel percentage for the entire column cross-section. The steel percentage is typically recommended to be maintained between 0.4% and 4% of the cross-sectional area of the column. However, exact values may be found in the National Annexes of each country. The calculation of minimum and maximum values is stated in the Eurocode 2 (EN 1992-1-1), Clause 9.5.2. Thereby, the normalized reinforcement index lines (I_{nr}) of the failure surface, also known as the boundary limit lines, of the 3D interaction surface or 2D interaction diagram should fall within these minimum (0.4%) and maximum (4%) recommended values. This ensures that the design remains within the prescribed limits.

There are three primary methods to determine the area of steel, each serving a different purpose, as follows.

6.1.3.1 Method 1: Steel Area for Constructing the N-M Digram Based on Minimum and Maximum Reinforcement Percentages

This method is utilized to construct the 2D interaction diagrams and 3D N-M interaction surface based on the maximum and minimum steel area limits stated in Eurocode 2 (EN 1992-1-1), Clause 9.5.2.

6.1.3.1.1 Method of Calculation

To proceed, the calculation is performed iteratively for specific steel percentages, chosen as needed (E.g., 4%). This is shown in Equations 6.22 and 6.23.

$$As_{Mb} = Sp_{Ratio} \cdot b \cdot h \tag{6.22}$$

$$Sp_{Ratio} = \frac{Sp_{Bc}}{100} \tag{6.23}$$

Where,
Sp_{Ratio}: Steel ratio for the steel bar in consideration (E.g., 0.01).
As_{Mb}: Maximum area allowed for the steel bar in consideration.
Sp_{Bc}: Steel percentage for the steel bar in consideration (E.g., 1%).
h: Height of the section.
b: Breadth of the section.

6.1.3.1.2 Example Calculation of Maximum Steel Area for Bar in Consideration (As_{Mb})

Given Data: Number of bars: 4, $\text{Sp}_{\text{Ratio}} = 4\%$, $h = 300\,\text{mm}$, $b = 225\,\text{mm}$.

Calculation:

Steel area for one bar of each corner $= \frac{4\%}{4} = 1\%$

So, maximum area of steel for the bar in consideration,

$$Sp_{Ratio} = \frac{Sp_{Bc}}{100}$$

$$Sp_{Ratio} = \frac{1}{100} = 0.01$$

$$As_{Mb} = \left(\frac{Sp_{Ratio}}{100}\right) \cdot b \cdot h$$

$$As_{Mb} = 0.01 \times 300\,\text{mm} \times 225\,\text{mm} = 675\,\text{mm}^2$$

6.1.3.2 Method 2: Steel Area for Constructing the N-M Digram with Custom Bar Parameters

This method is utilized when constructing the N-M interaction diagrams with actual bar parameters without explicitly specifying a reinforcement percentage. Nevertheless, it must still comply with the recommended minimum and maximum reinforcement limits stated in the Eurocode 2 (EN 1992-1-1), Clause 9.5.2.

6.1.3.2.1 Method of Calculation

The equation for the calculation of steel area is shown in Equation 6.24.

$$As_{Bc} = \frac{\pi d^2}{4} \leq As_{Mb} \tag{6.24}$$

Where:

As_{Bc}: Area of steel of the steel bar in consideration.
d = Diameter of the bar in consideration.
As_{Mb}: Maximum area allowed for the steel bar in consideration.

6.1.3.2.2 Example Calculation

Data: $As_{Mb} = 675\,\text{mm}^2$, $d = 20\,\text{mm}$, $h = 300\,\text{mm}$, $b = 225\,\text{mm}$.

Calculation:

$$
\begin{aligned}
As_{Bc} &= \frac{\pi d^2}{4} \leq As_{Mb} \\
&= \frac{3.141592654 \cdot (20\,\text{mm}^2)}{4} \\
&= 314.159\,\text{mm}^2
\end{aligned}
$$

6.1.3.3 Method 3: Normalized Reinforcement Index (I_{nr}) and the Area of Steel (A_s)

This method is utilized to construct the 2D interaction diagrams and 3D N-M interaction surface based on a dimensionless index. This is commonly referred to as the normalized reinforcement index (I_{nr}), is a key parameter in reinforced concrete design. It blends the influence of steel strength and area with the geometric and material properties of the concrete section.

6.1.3.3.1 Method of Calculation

This parameter is useful in two primary ways. The first is for constructing the interaction diagram by specifying a value for the normalized reinforcement index (e.g., 0.4 or 0.6). This is utilized to calculate the area of reinforcement required to generate the corresponding curve on the axial force versus moment interaction diagram or surface (N-M digram or surface). This calculation is repeated for different values of the index to develop a complete set of interaction curves.

The second way is that, if a design chart is already available and the moment (M) and axial force (N) are known, the appropriate reinforcement area required can be identified directly from the nearest curve that passes through or lies just above the demand point. This provides a fast and efficient way to estimate the required reinforcement without iterative calculations. This is shown in Equations 6.25 to 6.27.

$$I_{nr} = \frac{A_s \cdot f_{yk}}{bh \cdot f_{ck}} \tag{6.25}$$

$$A_s = \frac{I_{nr}}{\left(\frac{f_{yk}}{bh \cdot f_{ck}}\right)} \tag{6.26}$$

$$D = \sqrt{\frac{A_s \cdot 4}{\pi}} \tag{6.27}$$

Where,

I_{nr} = Normalized reinforcement index.

A_s = Area of reinforcement.

b: Breadth of the column cross-section.

h: Height of the column cross-section.

f_{yk}: Characteristic yield strength of steel (Typically 500 MPa or 500 N/mm^2).

f_{ck}: Characteristic compressive strength of concrete.

D: Diameter of the reinforcement bar in consideration.

π: pi (Typically 3.14159265359).

6.1.3.3.2 Example Calculation

Given Data:

h = 300 mm, b = 225 mm, fck = 30 N/mm^2, Number of bars = 4, I_{nr} = 0.4.

Calculation:

$$bh = 300\,\text{mm} \times 225\,\text{mm} = 67500\,\text{mm}^2$$

$$I_{nr} = \frac{A_s \cdot f_{yk}}{bh \cdot f_{ck}}$$

$$\begin{aligned}
A_s &= \frac{I_{nr}}{\left(\frac{f_{yk}}{bh \cdot f_{ck}}\right)} \\
&= \frac{0.4}{\left(\frac{500\,\mathrm{N/mm^2}}{300\,\mathrm{mm} \times 225\,\mathrm{mm} \times 30\,\mathrm{N/mm^2}}\right)} \\
&= \frac{0.4}{\left(\frac{500\,\mathrm{N/mm^2}}{2025000\,\mathrm{N}}\right)} \\
&= \frac{0.4}{0.00024691} \\
&= 1620\,\mathrm{mm^2}
\end{aligned}$$

$$\text{Area for each bar } (As_{\mathrm{bc}}) = \frac{1620\,\mathrm{mm^2}}{4\,\text{bars}} = 405\,\mathrm{mm^2}$$

$$\begin{aligned}
\text{Diameter of each reinforcement bar } (D) &= \sqrt{\frac{A_s \cdot 4}{\pi}} \\
&= \sqrt{\frac{405\,\mathrm{mm^2} \cdot 4}{3.14159265359}} \\
&= \sqrt{515.662\,\mathrm{mm^2}} \\
&= 22.708\,\mathrm{mm}
\end{aligned}$$

Note: In this case, to accurately represent the '$I_{nr} = 0.4$' line in the interaction diagrams or surface, you must use a reinforcement bar area of 405 mm^2 (corresponding to a bar diameter of 22.708 mm) in the stress (σyd_{bc}) calculations. However, for simplicity, we will use 314.159 mm^2 for the stress calculations (σyd_{bc}) in subsequent sections. This corresponds to a bar diameter of 20 mm, which is a size readily available in practical scenarios.

6.2 Calculation of Axial Force of Steel (N_s)

The resistive force in steel, also known as the axial force in steel reinforcement (N_s), is the force developed in the steel bars when they resist stresses in a reinforced concrete member. This force arises due to the ability of the steel reinforcement to withstand both tensile and compressive stresses under load, effectively balancing the internal forces within the concrete section. Since concrete has limited tensile capacity, the steel reinforcement compensates by providing the necessary tensile resistance.

6.2.1 Method of Calculation

This resistive axial force (N_s) can be calculated by summing up the resistive axial forces of each bar (Ns_{bc}). This is calculated by multiplying the design stress in steel (σsd_{bc}) of each bar, by the area of steel (A_s) of each bar. This is shown in the Equations 6.28 and 6.29. The Sigma (σ) is a Greek letter utilized for notation.

$$N_s = \sum_{k=1}^{n} Ns_{bc} \tag{6.28}$$

$$Ns_{bc} = \sigma yd_{bc} \cdot As_{bc} \tag{6.29}$$

Where,
N_s: Total resistant axial force of steel.
Ns_{bc}: Resistant axial force of each steel bar in consideration.
As_{bc}: Area of steel for each bar in consideration.
$\sum_{k=1}^{n}$: Indicates the summation of the resistive axial force for each bar from the first to the last bar.
k: Current number of each bar (eg: bar 1, bar 2 etc).
σyd_{bc}: Design stress in steel of bar in consideration (maximum allowable stress in each steel bar).

Note: Replace 'As_{bc}' in the Equation 6.29, with 'As_{Mb}' shown in Equation 6.22, when constructing the N-M interaction diagram or surface for the maximum (4%) and minimum (typically, 0.4%) limits specified in the Eurocode 2 or the National Annex. However, if you are constructing the N-M interaction diagram or surface for a custom-designed column, do not replace it and use 'As_{bc}' as it is. Moreover, if you are constructing N-M interaction diagram or surface for a specific I_{nr} line, for instance, $I_{nr} = 0.4$, then A_s obtained through I_{nr} calculation should be utilized.

6.2.1.1 Example Calculation

Given Data: $As_{bc} = 314.159\,\text{mm}^2$

Stress	Bar 1	Bar 2	Bar 3	Bar 4
σyd_{bc}	417.783 N/mm²	-326.667 N/mm²	-434.783 N/mm²	3.417 N/mm²

Calculation:

Axial force of each bar: $Ns_{bc} = \sigma yd_{bc} \cdot As_{bc}$

Bar 1: $Ns_{bc} = 417.783\,\text{N/mm}^2 \times 314.159\,\text{mm}^2 = 131,250.400\,\text{N}$

Bar 2: $Ns_{bc} = -326.667\,\text{N/mm}^2 \times 314.159\,\text{mm}^2 = -102,625.465\,\text{N}$

Bar 3: $Ns_{bc} = -434.783\,\text{N/mm}^2 \times 314.159\,\text{mm}^2 = -136,591.108\,\text{N}$

Bar 4: $Ns_{bc} = 3.417\,\text{N/mm}^2 \times 314.159\,\text{mm}^2 = 1,073.482\,\text{N}$

Total axial force of steel (N_s):

$$N_s = \sum N_{s,bc}$$

$$= (+131,250.400\,\text{N}) + (-102,625.465\,\text{N}) + (-136,591.108\,\text{N}) + (+1,073.482\,\text{N})$$

$$= -106,892.691\,\text{N} = -106.893\,\text{kN}$$

Note: The negative result indicates that the steel reinforcement is in net tension, meaning the tensile forces in the bars outweigh the compressive forces. However, this does not imply that the entire column is in tension, but rather that the section is predominantly experiencing tensile effects over compressive ones.

6.2.1.2 Summary of Calculations

Bar Number	Axial Force
Bar 1 (Ns,bc)	131,250.400 N
Bar 2 (Ns,bc)	-102,625.465 N
Bar 3 (Ns,bc)	-136,591.108 N
Bar 4 (Ns,bc)	1,073.482 N
Total (Ns)	-106.893 kN

6.3 Moment Calculation of Each Axis for Steel

In a reinforced concrete section subject to biaxial bending, the steel reinforcement develops internal moments that resist applied loads. These moments, referred to as resistive bending moments in steel, are also known as the moment capacity of steel. They represent the contributions of the steel reinforcement to resist bending moments about the x and y axes, respectively.

This section outlines the methods utilized to calculate moment resistance of steel reinforcement. Two different approaches are detailed, the non-inclined (un-rotated) axis method and the inclined (rotated) axis method. The following sections outline each method in detail.

6.3.1 Method 1 - Non-inclined (Un-rotated) Axis Method

This method calculates moments based on the non-inclined (un-rotated) principal axes of the section, which are typically aligned with the geometric x and y axes. The lever arms for steel reinforcement and concrete are measured perpendicular to these axes. These distances are then used to compute moment contributions from both the steel and concrete components.

6.3.1.1 Method of Calculation

The bending moments in steel along the non-inclined x-axis (M_{sxn}) and y-axis (M_{syn}) are determined as shown in the, Equations 6.30 and 6.33, and illustrated in Figure 6.2. The Sigma (Σ) is a Greek letter utilized for notation.

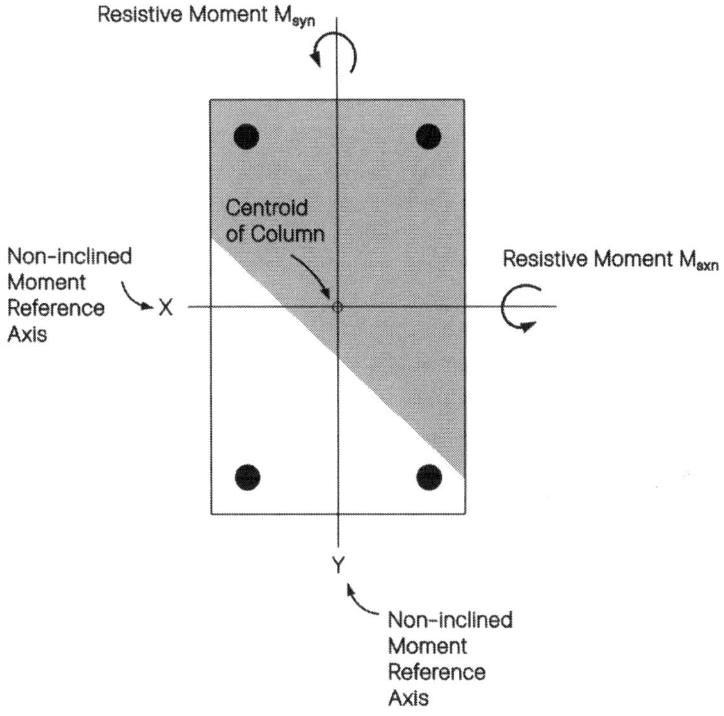

Figure 6.2: Resistive moments, M_{sxn} and M_{syn} in steel

$$M_{sxn} = \sum_{k=1}^{n} Msxn_{bc} \tag{6.30}$$

$$M_{syn} = \sum_{k=1}^{n} Msyn_{bc} \tag{6.31}$$

$$Msxn_{bc} = Ns_{bc} \cdot Lax_{mrs} \quad (6.32)$$

$$Msyn_{bc} = Ns_{bc} \cdot Lay_{mrs} \quad (6.33)$$

Where,
M_{sxn}: Resistive bending moment in steel along the non-inclined x-axis.
M_{syn}: Resistive bending moment in steel along the non-inclined y-axis.
$Msxn_{bc}$: Resistive bending moment of steel bar in consideration along the non-inclined x-axis.
$Msyn_{bc}$: Resistive bending moment of steel bar in consideration along the non-inclined y-axis.
Ns_{bc}: Resistive axial force of steel bar in consideration.
Lax_{mrs}: Lever arm for steel from non-inclined moment reference axis x to the centroid of steel reinforcement.
Lay_{mrs}: Lever arm for concrete from non-inclined moment reference axis y to the centroid of steel reinforcement.
Σ: Indicates sum of resistive moments of each bar.

6.3.1.2 Example Calculation of Moments (M_{sxn} and M_{syn})

The calculation example of the moments for each behaviour case is as follows.

Given Data:

Bar Number	**Axial Force(Ns_{bc})**	**Lax$_{mrs}$**	**Lay$_{mrs}$**
Bar 1	+131,250.400 N	+110 mm	+72.5 mm
Bar 2	-102,625.465 N	-110 mm	+72.5 mm
Bar 3	-136,591.108 N	-110 mm	-72.5 mm
Bar 4	+1,073.482 N	+110 mm	-72.5 mm

Calculation

Bar 1:

$$
\begin{aligned}
Msxn_{bc} &= Ns_{bc} \cdot Lax_{mrs} \\
&= 131{,}250.400\,\mathrm{N} \times 110\,\mathrm{mm} \\
&= 14{,}437{,}544\,\mathrm{Nmm} \\
&= 14{,}437{,}544\,\mathrm{Nmm} \times 10^{-6} \\
&= 14.438\,\mathrm{kNm}
\end{aligned}
$$

$$
\begin{aligned}
Msyn_{bc} &= Ns_{bc} \cdot Lay_{mrs} \\
&= 131{,}250.400\,\mathrm{N} \times 72.5\,\mathrm{mm} \\
&= 9{,}515{,}654\,\mathrm{Nmm} \\
&= 9{,}515{,}654\,\mathrm{Nmm} \times 10^{-6} \\
&= 9.516\,\mathrm{kNm}
\end{aligned}
$$

Bar 2:

$$
\begin{aligned}
Msxn_{bc} &= Ns_{bc} \cdot Lax_{mrs} \\
&= (-102{,}625.465\,\mathrm{N}) \times (-110\,\mathrm{mm}) \\
&= 11{,}288{,}801.15\,\mathrm{Nmm} \\
&= 11{,}288{,}801.15\,\mathrm{Nmm} \times 10^{-6} \\
&= 11.289\,\mathrm{kNm}
\end{aligned}
$$

$$
\begin{aligned}
Msyn_{bc} &= Ns_{bc} \cdot Lay_{mrs} \\
&= (-102{,}625.465\,\mathrm{N}) \times 72.5\,\mathrm{mm} \\
&= -7{,}440{,}346.213\,\mathrm{Nmm} \\
&= -7{,}440{,}346.213\,\mathrm{Nmm} \times 10^{-6} \\
&= -7.440\,\mathrm{kNm}
\end{aligned}
$$

Bar 3:

$$
Msxn_{bc} = Ns_{bc} \cdot Lax_{mrs}
$$

$$
\begin{aligned}
&= (-136{,}591.108\,\text{N}) \times (-110\,\text{mm}) \\
&= 15{,}025{,}021.88\,\text{Nmm} \\
&= 15{,}025{,}021.88\,\text{Nmm} \times 10^{-6} \\
&= 15.025\,\text{kNm}
\end{aligned}
$$

$$
\begin{aligned}
Msyn_{bc} &= Ns_{bc} \cdot Lay_{mrs} \\
&= (-136{,}591.108\,\text{N}) \times (-72.5\,\text{mm}) \\
&= 9{,}902{,}855.33\,\text{Nmm} \\
&= 9{,}902{,}855.33\,\text{Nmm} \times 10^{-6} \\
&= 9.903\,\text{kNm}
\end{aligned}
$$

Bar 4:

$$
\begin{aligned}
Msxn_{bc} &= Ns_{bc} \cdot Lax_{mrs} \\
&= 1{,}073.482\,\text{N} \times 110\,\text{mm} \\
&= 118{,}083.02\,\text{Nmm} \\
&= 118{,}083.02\,\text{Nmm} \times 10^{-6} \\
&= 0.118\,\text{kNm}
\end{aligned}
$$

$$
\begin{aligned}
Msyn_{bc} &= Ns_{bc} \cdot Lay_{mrs} \\
&= 1{,}073.482\,\text{N} \times (-72.5\,\text{mm}) \\
&= -77{,}827.445\,\text{Nmm} \\
&= -77{,}827.445\,\text{Nmm} \times 10^{-6} \\
&= -0.077\,\text{kNm}
\end{aligned}
$$

Total moments in each non-inclined (unrotated) axis

$$
M_{sxn} = \sum_{k=1}^{n} Msxn_{bc}
$$

$$
= 14.438\,\text{kNm} + 11.289\,\text{kNm} + 15.025\,\text{kNm} + 0.118\,\text{kNm}
$$

$$= 40.870\,\text{kNm}$$

$$M_{syn} = \sum_{k=1}^{n} Msyn_{bc}$$

$$= 9.516\,\text{kNm} + (-7.440\,\text{kNm}) + 9.903\,\text{kNm} + (-0.077\,\text{kNm})$$

$$= 11.902\,\text{kNm}$$

Note: Method 2 yields different results compared to Method 1, as it utilizes an inclined axis. Thereby, the moments are native to that inclined orientation. However, in a subsequent section, which calculates the resultant moments and their directions, both methods ultimately produce the same final result. This consistency arises because, despite the difference in axis inclination, both approaches are based on the same fundamental principles.

6.3.2 Method 2 - Inclined (Rotated) Axis Method

This method involves calculating moments using an inclined (rotated) moment reference axis. The inclined x axis is aligned parallel to the neutral axis of the section, while the inclined y axis is aligned perpendicular to the neutral axis.

6.3.2.1 Method of Calculation

In this method, the distances from the reinforcement centroids to the inclined (rotated) moment reference axis are measured as perpendicular distances (Lax_{mrsi} and Lay_{mrsi}) relative to the neutral axis. This approach directly considers the natural geometry of the neutral axis, ensuring that the moment contributions of reinforcement are calculated along the rotated coordinate system. Unlike the non-inclined method, which calculates distances in the x and y directions separately, this method evaluates the distances along the rotated axes.

Since this method aligns with the geometry of the neutral axis, some individuals may find it more convenient or intuitive to process compared to the non-inclined moment calculation. However, the inclined axis method involves additional steps, such as calculating the inclination angle and transforming distances. This is shown in the Equations 6.34 and 6.37 and illustrated in Figure 6.3. The Sigma (Σ) is a Greek letter utilized for notation.

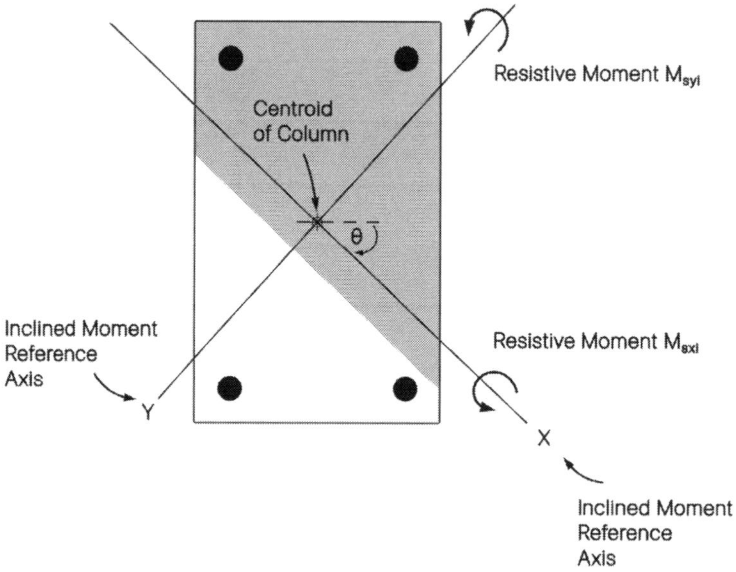

Figure 6.3: Resistive moments, M_{sxi} and M_{syi} in steel

$$M_{sxi} = \sum_{k=1}^{n} Msxi_{bc} \tag{6.34}$$

$$M_{syi} = \sum_{k=1}^{n} Msyi_{bc} \tag{6.35}$$

$$Msxi_{bc} = Ns_{bc} \cdot Lax_{mrs} \tag{6.36}$$

$$Msyi_{bc} = Ns_{bc} \cdot Lay_{mrs} \tag{6.37}$$

Where,

M_{sxi}: Total resistive bending moment in steel along the inclined x-axis.

M_{syi}: Total resistive bending moment in steel along the inclined x-axis.
$Msxi_{bc}$: Resistive bending moment of steel bar in consideration along the inclined x-axis.
$Msyi_{bc}$: Resistive bending moment of steel bar in consideration along the inclined y-axis.
Ns_{bc}: Resistive axial force of steel bar in consideration.
Lax_{mrsi}: Lever arm for steel from inclined moment reference axis x to the centroid of steel reinforcement.
Lay_{mrsi}: Lever arm for concrete from inclined moment reference axis y to the centroid of steel reinforcement.
Σ: Indicates sum of resistive moments of each bar.

6.3.2.2 Example Calculation of Moments (M_{sxi} and M_{syi})

The calculation example of the moments for each behaviour case is as follows.

Given Data:

Bar Number	Axial Force (Ns_{bc})	$\mathbf{Lax_{mrsi}}$	$\mathbf{Lay_{mrsi}}$
Bar 1	+131,250.400 N	+130.918 mm	-14.724 mm
Bar 2	-102,625.465 N	-38.091 mm	+126.116 mm
Bar 3	-136,591.108 N	-130.918 mm	+14.724 mm
Bar 4	+1,073.482 N	+38.091 mm	-126.116 mm

Calculation

Bar 1:

$$\begin{aligned} Msxi_{bc} &= Ns_{bc} \cdot Lax_{mrsi} \\ &= 131{,}250.400\,\text{N} \times 130.918\,\text{mm} \\ &= 17{,}183{,}039.87\,\text{Nmm} \\ &= 17{,}183{,}039.87\,\text{Nmm} \times 10^{-6} \end{aligned}$$

$$= 17.183\,\text{kNm}$$

$$\begin{aligned} Msyi_{bc} &= Ns_{bc} \cdot Lay_{mrsi} \\ &= 131{,}250.400\,\text{N} \times (-14.724)\,\text{mm} \\ &= -1{,}932{,}530.890\,\text{Nmm} \\ &= -1{,}932{,}530.890\,\text{Nmm} \times 10^{-6} \\ &= -1.933\,\text{kNm} \end{aligned}$$

Bar 2:

$$\begin{aligned} Msxi_{bc} &= Ns_{bc} \cdot Lax_{mrsi} \\ &= (-102{,}625.465\,\text{N}) \times (-38.091)\,\text{mm} \\ &= 3{,}909{,}106.587\,\text{Nmm} \\ &= 3{,}909{,}106.587\,\text{Nmm} \times 10^{-6} \\ &= 3.909\,\text{kNm} \end{aligned}$$

$$\begin{aligned} Msyi_{bc} &= Ns_{bc} \cdot Lay_{mrsi} \\ &= (-102{,}625.465\,\text{N}) \times 126.116\,\text{mm} \\ &= -12{,}942{,}713.14\,\text{Nmm} \\ &= -12{,}942{,}713.14\,\text{Nmm} \times 10^{-6} \\ &= -12.943\,\text{kNm} \end{aligned}$$

Bar 3:

$$\begin{aligned} Msxi_{bc} &= Ns_{bc} \cdot Lax_{mrsi} \\ &= (-136{,}591.108\,\text{N}) \times (-130.918\,\text{mm}) \\ &= 17{,}882{,}234.68\,\text{Nmm} \\ &= 17{,}882{,}234.68\,\text{Nmm} \times 10^{-6} \\ &= 17.882\,\text{kNm} \end{aligned}$$

$$\begin{aligned}
Msyi_{bc} &= Ns_{bc} \cdot Lay_{mrsi} \\
&= (-136{,}591.108\,\mathrm{N}) \times 14.724\,\mathrm{mm} \\
&= -2{,}011{,}167.474\,\mathrm{Nmm} \\
&= -2{,}011{,}167.474\,\mathrm{Nmm} \times 10^{-6} \\
&= -2.011\,\mathrm{kNm}
\end{aligned}$$

Bar 4:

$$\begin{aligned}
Msxi_{bc} &= Ns_{bc} \cdot Lax_{mrsi} \\
&= 1{,}073.482\,\mathrm{N} \times 38.091\,\mathrm{mm} \\
&= 40{,}890.003\,\mathrm{Nmm} \\
&= 40{,}890.003\,\mathrm{Nmm} \times 10^{-6} \\
&= 0.041\,\mathrm{kNm}
\end{aligned}$$

$$\begin{aligned}
Msyi_{bc} &= Ns_{bc} \cdot Lay_{mrsi} \\
&= 1{,}073.482\,\mathrm{N} \times (-126.116\,\mathrm{mm}) \\
&= -135{,}383.256\,\mathrm{Nmm} \\
&= -135{,}383.256\,\mathrm{Nmm} \times 10^{-6} \\
&= -0.135\,\mathrm{kNm}
\end{aligned}$$

Total moments in each inclined axis:

$$\begin{aligned}
Msxi &= \sum_{k=1}^{n} Msxi_{bc} \\
&= 17{,}183{,}039.87\,\mathrm{Nmm} + 3{,}909{,}106.587\,\mathrm{Nmm} + \\
&\quad 17{,}882{,}234.68\,\mathrm{Nmm} + 40{,}890.003\,\mathrm{Nmm} \\
&= 39{,}015{,}271.14\,\mathrm{Nmm} \\
&= 39{,}015{,}271.14 \times 10^{-6} \\
&= 39.015\,\mathrm{kNm}
\end{aligned}$$

$$Msyi = \sum_{k=1}^{n} Msyi_{bc}$$

$$= (-1{,}932{,}530.890\,\text{Nmm}) + (-12{,}942{,}713.14\,\text{Nmm}) + (-2{,}011{,}167.474\,\text{Nmm}) + (-135{,}383.256\,\text{Nmm})$$

$$= -17{,}021{,}794.76\,\text{Nmm}$$

$$= -17{,}021{,}794.76 \times 10^{-6}$$

$$= -17.022\,\text{kNm}$$

Note: Method 2 yields different results compared to Method 1, as it utilizes an inclined axis. Thereby, the moments are native to that inclined orientation. However, in a subsequent section, which calculates the resultant moments and their directions, both methods ultimately produce the same final result. This consistency arises because, despite the difference in axis inclination, both approaches are based on the same fundamental principles.

6.4 Summation of the Axial Forces and Moments of Steel and Concrete for Each Axis

The total axial force (N), and total resistive moments along x axis (M_x) and y axis (M_y), are determined through the summation as follows.

6.4.1 Non-inclined (Unrotated) Axes

The summation of the forces along the non-inclined axes is as follows.

6.4.1.1 Method of Calculation

The Equation 6.38 shows the summation of axial forces of concrete (N_c) and the steel reinforcement (N_s). The Equations 6.39 and 6.40 show the summation of the moments of concrete (M_{cxn} and M_{cyn}) and the steel reinforcement (M_{sxn} and M_{syn}), which utilize the non-inclined moment reference axes.

$$N = N_c + Ns \tag{6.38}$$

$$M_x = M_{cxn} + M_{sxn} \tag{6.39}$$

$$M_y = M_{cyn} + M_{syn} \tag{6.40}$$

Where,

N_c: Resistive axial compressive force in concrete.
M_x: Total resistive bending moment along the non-inclined x-axis.
M_y: Total resistive bending moment along the non-inclined y-axis.
M_{cxn}: Resistive bending moment in concrete along the non-inclined x-axis.
M_{cyn}: Resistive bending moment in concrete along the non-inclined y-axis.
M_{sxn}: Resistive bending moment in steel along the non-inclined x-axis.
M_{syn}: Resistive bending moment in steel along the non-inclined y-axis.

6.4.1.2 Example Calculation

Given Data:

$Nc = 235.008\,\text{kN}$, $Ns = -106.893\,\text{kN}$, $M_{cxn} = 22.718\,\text{kNm}$,
$M_{cyn} = 11.398\,\text{kNm}$, $M_{sxn} = 40.869\,\text{kNm}$, $M_{syn} = 11.902\,\text{kNm}$

$$\begin{aligned}\text{Total Axial Force:}\quad N &= Nc + Ns\\ &= 235.008\,\text{kN} + (-106.893\,\text{kN})\\ &= 128.115\,\text{kN}\end{aligned}$$

$$\begin{aligned}\text{Total Moment in } x\text{-axis:}\quad Mx &= Mcxn + Msxn\\ &= 22.718\,\text{kNm} + 40.869\,\text{kNm}\\ &= 63.587\,\text{kNm}\end{aligned}$$

$$\begin{aligned}\text{Total Moment in } y\text{-axis:}\quad My &= Mcyn + Msyn\\ &= 11.398\,\text{kNm} + 11.902\,\text{kNm}\\ &= 23.300\,\text{kNm}\end{aligned}$$

6.4.2 Inclined (Rotated) Axes

The summation of the forces along the inclined axes is as follows.

6.4.2.1 Method of Calculation

The Equation 6.41 shows the summation of axial forces of concrete (N_c) and the steel reinforcement (N_s). The Equations 6.42 and 6.43 show the summation of the moments of concrete (M_{cxi} and M_{cyi}) and the steel reinforcement (M_{sxi} and M_{syi}), which utilize the inclined moment reference axes.

$$N = N_c + N_s \tag{6.41}$$

$$M_x = M_{cxi} + M_{sxi} \tag{6.42}$$

$$M_y = M_{cyi} + M_{syi} \tag{6.43}$$

Where,
N_c: Resistive axial compressive force in concrete.
M_x: Total resistive bending moment along the inclined x-axis.
M_y: Total resistive bending moment along the inclined y-axis.
M_{cxi}: Resistive bending moment in concrete along the inclined x-axis.
M_{cyi}: Resistive bending moment in concrete along the inclined y-axis.
M_{sxi}: Resistive bending moment in steel along the inclined x-axis.
M_{syi}: Resistive bending moment in steel along the inclined y-axis.

6.4.2.2 Example Calculation

Given Data:
$N_c = 235.008\,\text{kN}$, $N_s = -106.893\,\text{kN}$, $M_{cxi} = 24.749\,\text{kNm}$,
$M_{cyi} = -5.788\,\text{kNm}$, $M_{sxi} = 39.015\,\text{kNm}$, $M_{syi} = -17.022\,\text{kNm}$

$$\begin{aligned}\text{Total Axial Force:}\quad N &= N_c + N_s\\ &= 235.008\,\text{kN} + (-106.893\,\text{kN})\end{aligned}$$

$$= 128.115\,\text{kN}$$

$$\text{Total Moment in } x\text{-axis:} \quad M_x = M_{cxi} + M_{sxi}$$
$$= 24.749\,\text{kNm} + 39.015\,\text{kNm}$$
$$= 63.764\,\text{kNm}$$

$$\text{Total Moment in } y\text{-axis:} \quad M_y = M_{cyi} + M_{syi}$$
$$= (-5.788\,\text{kNm}) + (-17.022\,\text{kNm})$$
$$= -22.810\,\text{kNm}$$

This page intentionally left blank.

Chapter 7:

Resultant Moment and Direction of Steel and Concrete

The resultant demand bending moment (M_{td}) is the external bending moment acting on the column, which must be resisted to maintain stability. It can originate from connected beams, slabs, or other applied loads. The resultant resistive bending moment (M_{tr}) is the internal moment generated by the concrete and reinforcement in the column to counteract the demand moment. For structural stability, the resistive moment must be greater than or equal to the demand moment.

In a biaxial bending scenario, a column is subjected to simultaneous moments about the x and y axes. The true direction of bending is determined by the resultant demand bending moment, which generally does not align or remain perpendicular to the neutral axis (NA).

The resultant moment and direction of the steel reinforcement and concrete are described as follows.

7.1 Resultant Moment of Steel and Concrete

The resultant moment is a combined measure of the net bending effect on the column when moments act simultaneously about both the x and y axes. It represents the maximum bending resistance in the resultant resistive bending direction.

7.1.1 Method of Calculation

The calculation formula for the resultant moment is shown in Equations 7.1 and 7.2. This applies the Pythagorean theorem to combine the orthogonal moment components. This is further illustrated in Figures 7.1 and 7.2.

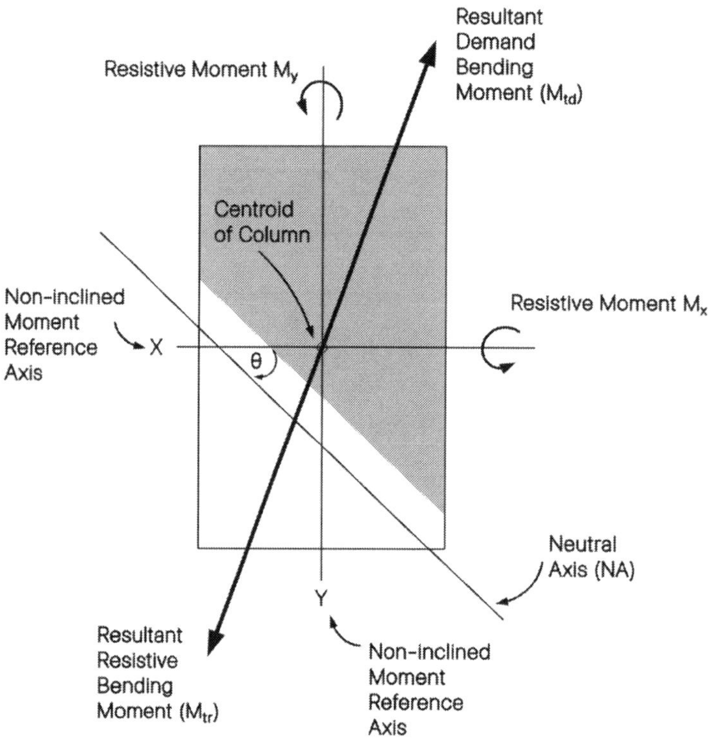

Figure 7.1: Total resistive and demand moments with non-inclined x and y axes

$$M_{tr} = \sqrt{M_x^2 + M_y^2} \tag{7.1}$$

$$M_{td} = M_{tr} \tag{7.2}$$

Where,

M_{tr}: Total resultant resistive bending moment.

M_{td}: Total resultant demand bending moment.

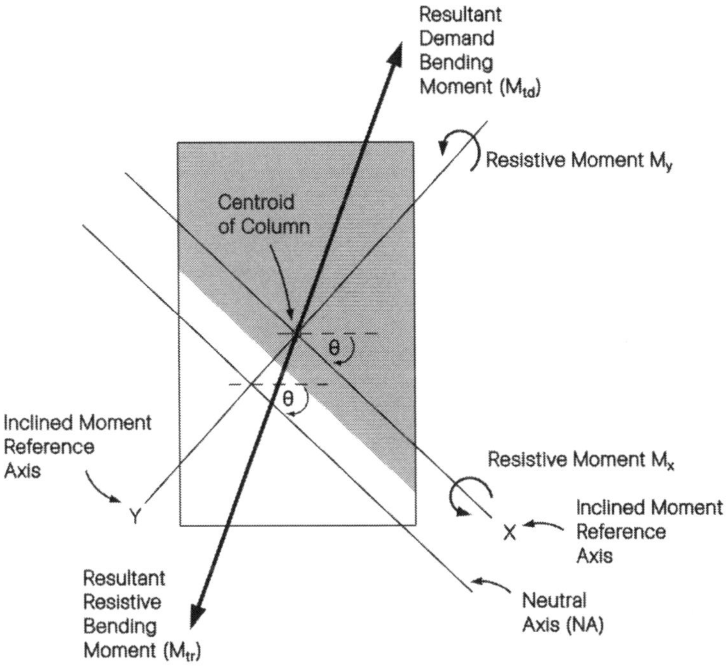

Figure 7.2: Total resistive and demand moments with inclined x and y axes

M_x: Total resistive bending moment along the non-inclined or inclined x-axis.
M_y: Total resistive bending moment along the non-inclined or inclined y-axis.

Note 1: Moments may be used from either inclined or non-inclined axes. Nevertheless, ensure consistency by not mixing the two types. For instance, if inclined moments are used for M_x, then M_y should also be from inclined moments, and vice versa.

Note 2: The total resultant resistive bending moment (M_{tr}) should always be equal to or greater than the total resultant demand bending moment (M_{td}) to ensure structural safety. In this case, $M_{tr} = M_{td}$ means the section is at its limit equilibrium with the applied moments.

7.1.2 Example Calculation

For the non-inclined axes

Given Data: $M_x = 63.587\,\text{kNm}, \quad M_y = 23.3\,\text{kNm}$

$$
\begin{aligned}
M_{tr} &= \sqrt{M_x^2 + M_y^2} \\
&= \sqrt{(63.587\,\text{kNm})^2 + (23.3\,\text{kNm})^2} \\
&= 67.721\,\text{kNm}
\end{aligned}
$$

For the inclined axes

Given Data: $M_x = 63.764\,\text{kNm}, \quad M_y = -22.810\,\text{kNm}$

$$
\begin{aligned}
M_{tr} &= \sqrt{M_x^2 + M_y^2} \\
&= \sqrt{(63.764\,\text{kNm})^2 + (-22.810\,\text{kNm})^2} \\
&= 67.721\,\text{kNm}
\end{aligned}
$$

Note 3: Although the M_x and M_y values differ depending on whether the axes are non-inclined and inclined, the resultant moment remains the same. This is because the physical behaviour of the member does not depend on the choice of coordinate axes.

Note 4: The resultant moment is less than the direct sum of M_x and M_y. For instance, adding 63.587 kNm + 23.3 kNm without using the equation, results in 86.887 kNm. That is higher than the resultant moment of 67.721 kNm. That means the column "feels" less force in the resultant bending direction compared to the combined resistance required when treating the moments separately along the x and y axes.

7.2 Direction of the Resultant Resistive and Demand Moments for Steel and Concrete

The angle of direction of the resultant resistive and demand moments (θM_{xn} and θMd_{xn}) to the x-axis utilizes '$atan2$', which is a trigonometric function that calculates the arctangent of the ratio 'M_y/M_x', while considering the signs of both M_y and M_x to determine the correct quadrant of the angle. The '$atan2$' function accounts for the quadrant of the vector, thereby avoiding ambiguity in the angle calculation. It should not be confused with the basic $atan$ function, which only returns values between $-\pi/2$ and $\pi/2$, and does not account for the quadrant of the vector.

Note 1: Moments may be used from either inclined or non-inclined axes. Nevertheless, ensure consistency by not mixing the two types. If inclined moments are used for Mx, then My should also be from inclined moments, and vice versa.

Note 2: When utilizing spreadsheet software, ensure that the output of the atan2 function is converted to degrees by using '$= DEGREES(ATAN2(M_y, M_x))$', as some software returns '$atan2$' values in radians by default.

7.2.1 Non-inclined (Unrotated) Axes

The method and an example calculation for the direction of the resultant moment utilizing non-inclined axes are as follows.

7.2.1.1 Method of Calculation

The angle of direction from the resultant resistive and demand moments from non-inclined moment reference axis x and the neutral axis are shown Equation 7.3 to Equation 7.6. Moreover, this is illustrated in

Figure 7.3.

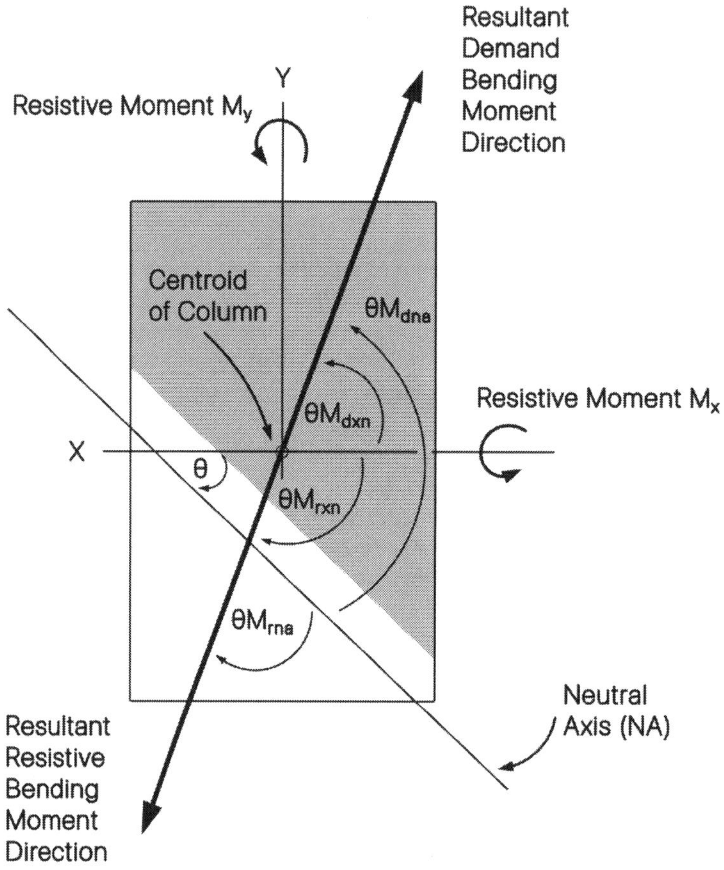

Figure 7.3: Resultant resistive and demand moment direction angles from non-inclined axis

$$\theta Mr_{xn} = 180^\circ - \text{atan2}(M_y, M_x) \tag{7.3}$$

$$\theta Mr_{na} = 180^\circ - (\theta Md_{xn} + \theta) \tag{7.4}$$

$$\theta Md_{xn} = \text{atan2}(M_y, M_x) \tag{7.5}$$

$$\theta Md_{na} = \theta Md_{xn} + \theta \tag{7.6}$$

Where,

θMr_{xn}: Angle from non-inclined (unrotated) x-axis to the resultant resistive moment direction.

θMr_{na}: Angle from neutral axis to the resultant resistive moment direction.
θMd_{xn}: Angle from non-inclined (unrotated) x-axis to the resultant demand moment direction.
θMd_{na}: Angle from neutral axis to the resultant demand moment direction.
θ: Inclination angle (rotation angle) of the inclined (rotated) axis relative to the non-inclined (unrotated) axis.
$atan2(M_y, M_x)$: A trigonometric function that calculates the arctangent of the ratio M_y/M_x.
M_x: Total moment along the non-inclined x-axis.
M_y: Total moment along the non-inclined y-axis.

7.2.1.2 Example Calculation

Given Data: $Mx = 63.587\,\text{kNm}, \quad My = 23.298\,\text{kNm}, \quad \theta = 39.805571092^\circ$

Angle from non-inclined (unrotated) x-axis to the resultant resistive moment direction:

$$\begin{aligned}\theta Mr_{xn} &= 180^\circ - atan2(My, Mx)\\ &= 180^\circ - atan2(23.298\,\text{kNm}, 63.587\,\text{kNm})\\ &= 110.123^\circ\end{aligned}$$

Angle from neutral axis to the resultant resistive moment direction:

$$\begin{aligned}\theta Mr_{na} &= 180^\circ - (\theta Md_{xn} + \theta)\\ &= 180^\circ - (69.87731176^\circ + 39.805571092^\circ)\\ &= 70.317^\circ\end{aligned}$$

Angle from non-inclined (unrotated) x-axis to the resultant demand moment direction:

$$\begin{aligned}\theta Md_{xn} &= atan2(My, Mx)\\ &= atan2(23.298\,\mathrm{kNm}, 63.587\,\mathrm{kNm})\\ &= 69.877^\circ\end{aligned}$$

Angle from neutral axis to the resultant demand moment direction:

$$\begin{aligned}\theta Md_{na} &= \theta Md_{xn} + \theta\\ &= 69.877^\circ + 39.805571092^\circ\\ &= 109.683^\circ\end{aligned}$$

7.2.2 Inclined (Rotated) Axes

The method and an example calculation for the direction of the resultant moment utilizing inclined axes are as follows.

7.2.2.1 Method of Calculation

The angle of direction from the resultant resistive and demand moments from inclined moment reference axis x and the neutral axis are shown in Equation 7.7 to Equation 7.11. Moreover, the angle of direction from non-inclined x axis can also be determined utilizing these equations, as shown in Equations 7.12 and 7.13. These are illustrated in figure 7.4.

$$\theta Mr_{xi} = 180^\circ - \mathrm{atan2}(M_y, M_x) \tag{7.7}$$

$$\theta Mr_{na} = 180^\circ - \theta Md_{xi} \tag{7.8}$$

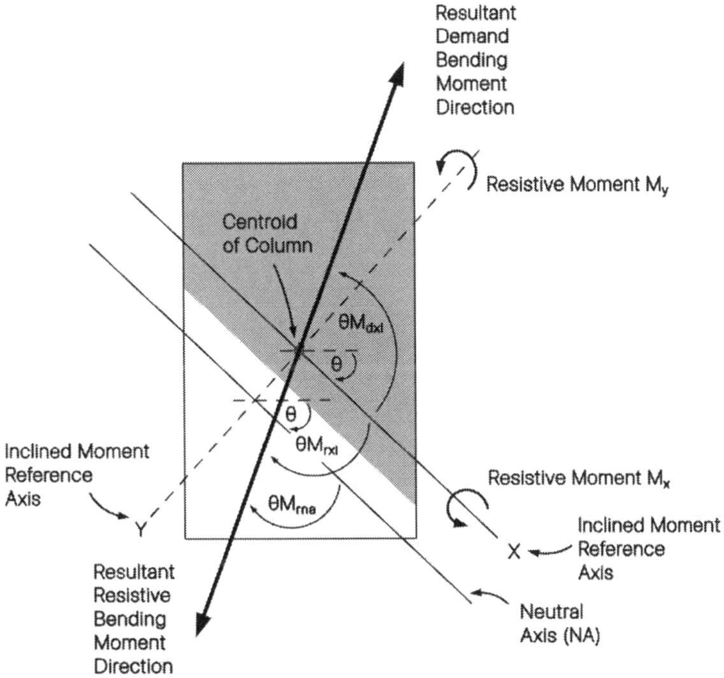

Figure 7.4: Resultant resistive and demand moment direction angles from inclined axis

$$\theta Mr_{na} = \theta Mr_{xi} \tag{7.9}$$

$$\theta Md_{xi} = \text{atan2}(M_y, M_x) \tag{7.10}$$

$$\theta Md_{na} = \theta Md_{xi} \tag{7.11}$$

$$\theta Mr_{xn} = 180^\circ - (\theta Md_{xi} - \theta) \tag{7.12}$$

$$\theta Md_{xn} = \theta Md_{xi} - \theta \tag{7.13}$$

Where,
θMr_{xi}: Angle from inclined (rotated) x-axis to the resultant resistive moment direction.
θMr_{na}: Angle from neutral axis to the resultant resistive moment direction.
θMr_{xn}: Angle from non-inclined (unrotated) x-axis to the resultant resistive moment direction.
θMd_{xi}: Angle from inclined (rotated) x-axis to the resultant demand

moment direction.
θMd_{na}: Angle from neutral axis to the resultant demand moment direction.
θMd_{xn}: Angle from non-inclined (unrotated) x-axis to the resultant demand moment direction.
θ: Inclination angle (rotation angle) of the inclined (rotated) axis relative to the non-inclined (unrotated) axis.
atan2(M_y, M_x): A trigonometric function that calculates the arctangent of the ratio M_y/M_x.
M_x: Total moment along the inclined x-axis.
M_y: Total moment along the inclined y-axis.

7.2.2.2 Example Calculation

Given Data: $Mx = 63.764\,\text{kNm}, \quad My = -22.810\,\text{kNm}, \quad \theta = 39.805571092°$

Angle from inclined (rotated) x-axis to the resultant resistive moment direction:

$$\begin{aligned}\theta Mr_{xi} &= 180° - atan2(My, Mx)\\ &= 180° - atan2(-22.810\,\text{kNm}, 63.764\,\text{kNm})\\ &= 70.317°\end{aligned}$$

Angle from inclined (rotated) x-axis to the resultant moment direction:

$$\begin{aligned}\theta Md_{xi} &= atan2(My, Mx)\\ &= atan2(-22.810\,\text{kNm}, 63.764\,\text{kNm})\\ &= 109.683°\end{aligned}$$

Angle from neutral axis to the resultant resistive moment direction:

$$\theta Mr_{na} = 180° - \theta Md_{xi}$$
$$= 180° - 109.683°$$
$$= 70.317°$$

Since $\theta Mr_{na} = \theta Mr_{xi}$, it can be confirmed that: $\theta Mr_{xi} = 70.317°$

Angle from neutral axis to the resultant moment direction:

$$\theta Md_{na} = \theta Md_{xi}$$
$$= 109.682°$$

Angle from non-inclined (unrotated) x-axis to the resultant resistive moment direction:

$$\theta Mr_{xn} = 180° - (\theta Md_{xi} - \theta)$$
$$= 180° - (109.682° - 39.805571092°)$$
$$= 110.124°$$

Utilizing inclined (rotated) axis data to obtain the angle from non-inclined (unrotated) x-axis to the resultant moment direction:

$$\theta Md_{xn} = \theta Md_{xi} - \theta$$

$$= 109.682° - 39.805571092°$$

$$= 69.876°$$

Chapter 8:

Plotting the Two-dimensional (2D) Interaction Diagrams

The construction of the 2D interaction diagrams follows a systematic approach that integrates the calculations of concrete and steel contributions, moment capacities, and axial force effects. The following sections provide a detailed breakdown of the formulation of N-M interaction relationships, which offers a comprehensive representation of axial and moment capacity of the column under various loading conditions.

8.1 Reinforcement Depth Ratio (R_{dr})

The reinforcement depth ratio R_{dr} is a normalized measure of the vertical position of the reinforcement within a concrete section. It is defined as the ratio between the effective cover depth d' and the total section height h. This parameter normalizes the reinforcement depth relative to the section height, aiding in comparative design. However it does not uniquely represent the quantity or specific arrangement of reinforcement within the section.

8.1.1 Method of Calculation

The value of R_{dr} is calculated using the simple ratio $\frac{d'}{h}$, where d' is the effective cover, defined as the distance from the extreme compression fibre to the centroid of the compression reinforcement, and h is the overall height of the cross-section. This is shown in Equation 8.1 and illustrated in Figure 8.1.

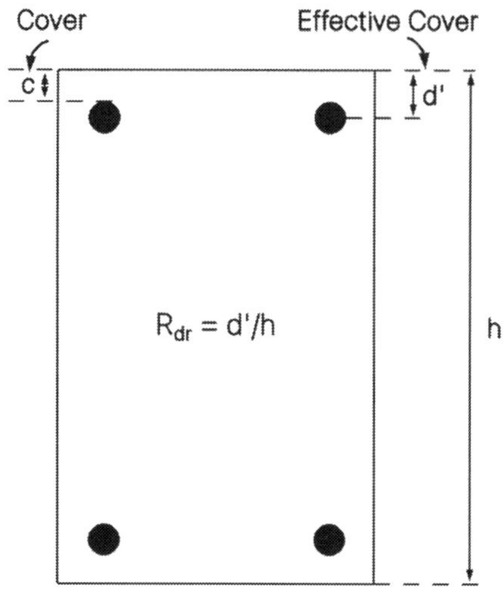

Figure 8.1: Illustration of the reinforcement depth ratio (R_{dr})

$$R_{dr} = \frac{d'}{h} \tag{8.1}$$

Where,

R_{dr}: Reinforcement depth ratio.

d': Effective cover

h: Height of the section.

8.1.2 Example Calculation

Given Data: $d' = 40$ mm, $h = 300$ mm

$$\begin{aligned} R_{dr} &= \frac{d'}{h} \\ &= \frac{40mm}{300mm} \\ &= 0.133 \end{aligned}$$

8.2 Applied Moment to Maximum Moment Ratio (R_m)

The parameter R_m represents the ratio of the applied bending moment to the maximum moment capacity of the column cross-section at a given axial load. It quantifies how much of the bending capacity of the section is being utilized under a specific loading condition.

8.2.1 Method of Calculation

A value of '$R_m = 0$' corresponds to a purely axial loading condition with no bending, while '$R_m = 1$' indicates that the applied moment equals the full moment resistance of the section at that axial force level, representing a fully utilized bending condition.

On an M-N interaction diagram, the line defined by '$R_m = 1$' is simply the straight line connecting the maximum moment points of each axial load curve. This line forms the upper boundary of moment capacity across the entire axial load range. A simplified illustration of this is shown in Figure 8.2, where two curves represent interaction limits at different axial loads, and the dashed line joining their respective maximum moment points marks the '$R_m = 1$' reference. The rest of the diagram region can then be subdivided proportionally to visualize intermediate 'R_m' values such as 0.1, 0.2, and so on, up to 1, allowing a clear assessment of moment utilization throughout the interaction domain. This is shown in Equations 8.2 and 8.3.

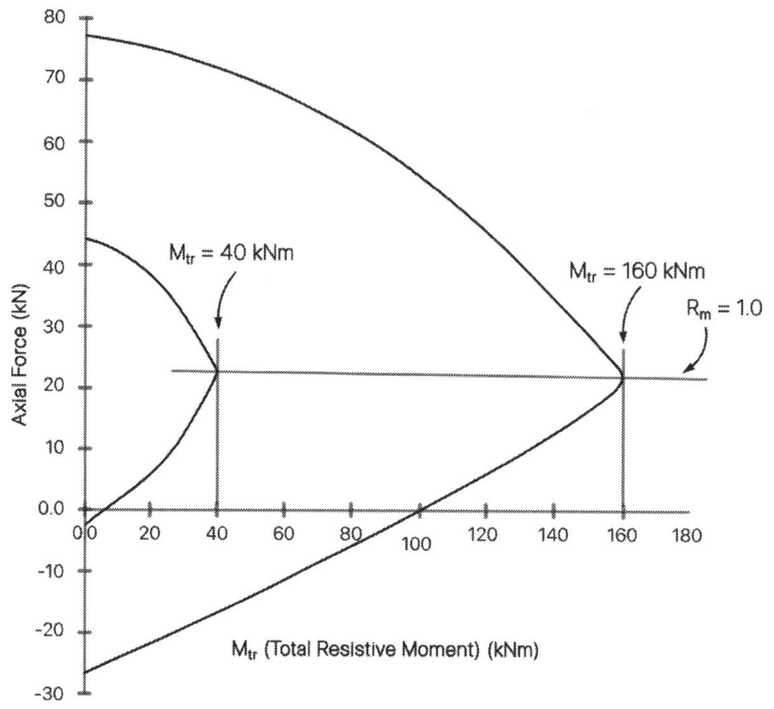

Figure 8.2: Illustration of the reference line corresponding to $R_m = 1.0$

$$R_m = \frac{M_a}{M_{tr}} \tag{8.2}$$

$$M_a = R_m \cdot M_{tr} \tag{8.3}$$

Where,
R_m: The ratio of the applied bending moment to the maximum moment capacity of the column cross-section at a given axial load.
M_a: Applied moment.
M_{tr}: Total resultant resistive moment.

8.2.2 Example Calculation

Given Data: Curve 1: $M_{tr} = 160$ kNm, Curve 2: $M_{tr} = 40$ kNm, $R_m = 0.4$

$$M_a = R_m \cdot M_{tr}$$

$$= 0.4 \times 160 \text{ kNm}$$

$$= 64 \text{ kNm}$$

$$M_a = R_m \cdot M_{tr}$$

$$= 0.4 \times 40 \text{ kNm}$$

$$= 16 \text{ kNm}$$

The Figure 8.3 illustrates R_m reference line resulting from the calculation.

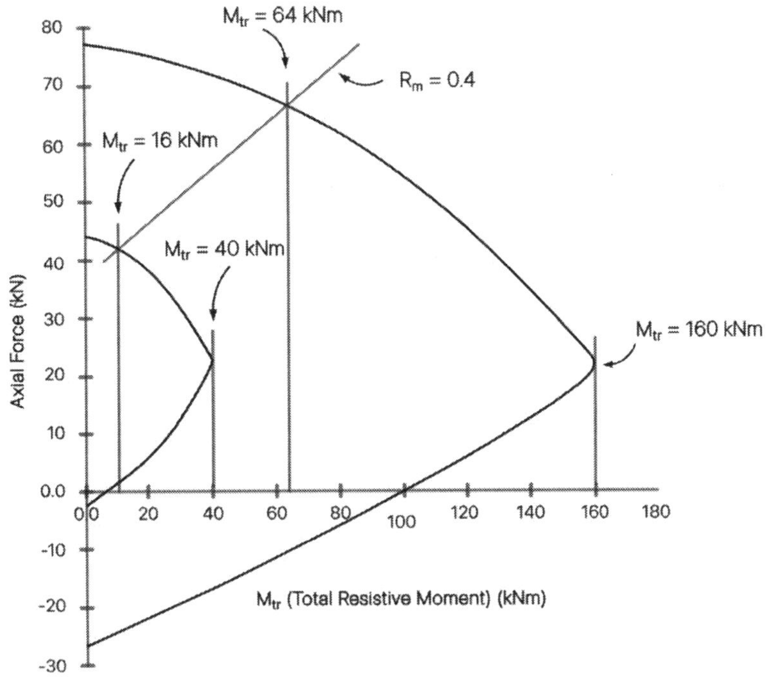

Figure 8.3: Illustration of the example calculation of $R_m = 0.4$

8.3 Two-dimensional (2D) Interaction Diagrams of Axial Force (N) and Total Moments (M_{tr})

The two-dimensional (2D) interaction diagram, also known as the interaction graph or curve, is constructed using data of axial force (N), total resistive moment (M_{tr}). The inclination angle (θ) of the neutral axis is set to 0° for the x-axis diagram and 90° for the y-axis diagram. This is shown in Figures 8.5 and 8.6. Moreover, the diagram can also be constructed for custom angles, for instance, Figure 8.7 shows the interaction diagram for the angle of inclination (θ) of 39.805571092°, for x and y axis respectively.

The reduction factor ($\omega = 0.9$) utilized in previous sections does not apply to 90° and 0° inclination angles. This factor accounts for the nonlinear distribution of the compression zone when it decreases toward the extreme compression fibre. Eurocode 2, Clause 3.1.7 (3) states this condition. However, in this case, the neutral axis is set parallel to the x-axis or y-axis. As a result, the extreme compression fibre maintains its full breadth (b) or height (h). Therefore, the clause is not relevant in this context. Consequently, when calculating the moments and forces to construct the 90° and 0° interaction diagrams, the reduction factor (ω) of 0.9 should not be applied. However, if any other angle besides 0° or 90° is used, the 0.9 reduction factor (ω) should be applied.

The neutral axis started at the extreme compression fibre intersection point (ECFIP). It is then shifted gradually until the total resultant resistive moment (M_{tr}) becomes zero or almost zero. Consequently, the data were obtained. The shifting of neutral axis (NA) is shown in Figure 8.4. Moreover, the neutral axis distance along the height (δh) and the neutral axis distance along the breadth (δb) are maintained at a preferred inclination ratio (θ) of choice. This ensures that the neutral axis inclination angle remains constant.

Creating the interaction diagrams with smaller increments, that is, adding more data points, reduces the inherent conservatism of the curves and yields interaction diagrams that are correspondingly less conservative.

The N-M interaction diagrams shown in Figures 8.5 and 8.6 were developed utilizing $f_{\mathrm{cd}} = 17\,\mathrm{N/mm^2}$ (derived from $f_{\mathrm{ck}} = 30\,\mathrm{N/mm^2}$), four 20

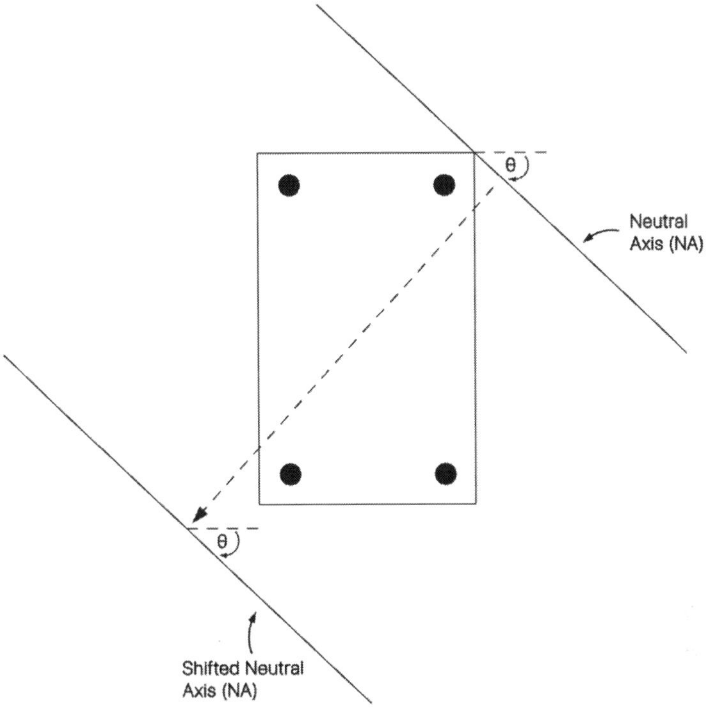

Figure 8.4: Neutral axis shift

mm diameter bars, a cover (c) of 30 mm (effective cover (d') = 40 mm), $h = 300\,\text{mm}$, $b = 225\,\text{mm}$, Normalized reinforcement index (I_{nr}) = 0.4, Reinforcement depth ratio (R_{dr}) = 0.133, and the reduction factor $\omega = 1$. Figure 8.7 utilized an angle of inclination (θ) of 39.805571092°, which lies between 90° and 0°. Accordingly, the reduction factor $\omega = 0.9$ was utilized.

The data utilized to plot these diagrams are provided in Tables 9.1, 9.2 and 9.3 in Annexture. Moreover, the data tables provided in the annexture, along with the charts shown in this section, can also be downloaded by visiting the website www.WattagePublishers.com.

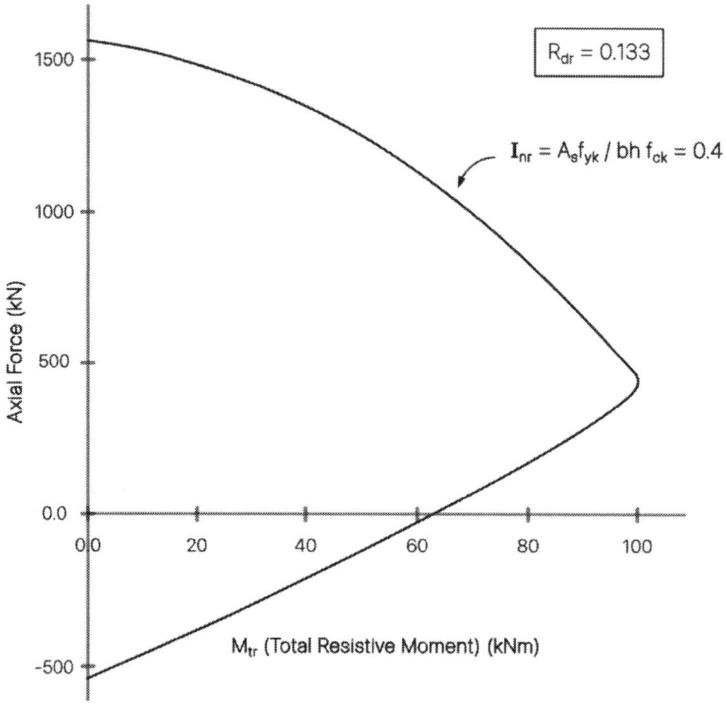

Figure 8.5: N-M Interaction diagram at $\theta = 0°$ (for x axis), $\omega = 1.0$ and $R_{dr} = 0.133$

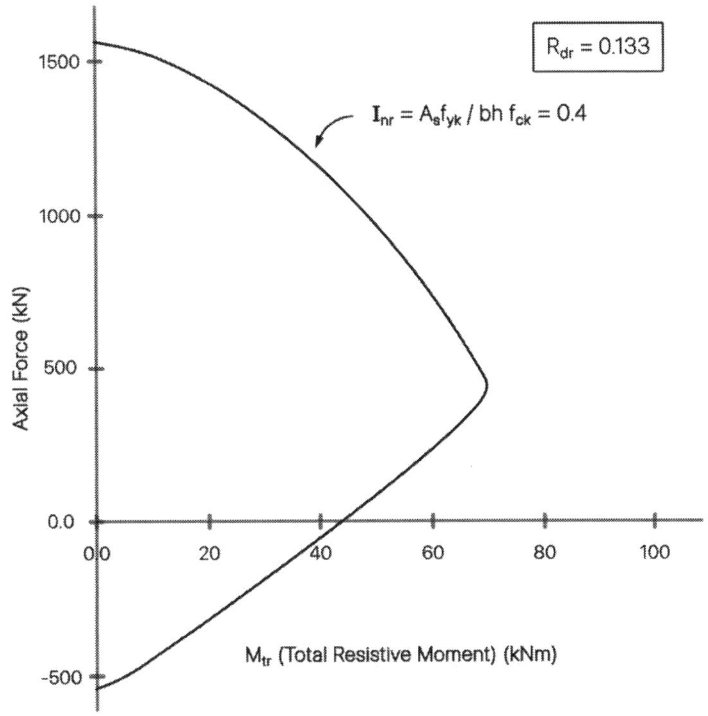

Figure 8.6: N-M Interaction diagram at $\theta = 90°$ (for y axis), $\omega = 1.0$ and $R_{dr} = 0.133$

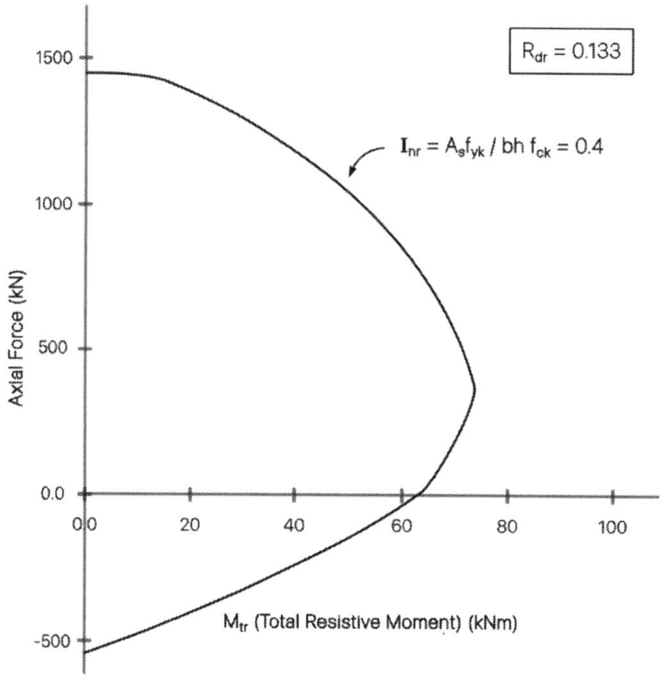

Figure 8.7: N-M Interaction diagram at $\theta = 39.805571092°$, $\omega = 0.9$ and $R_{dr} = 0.133$

8.4 Normalized Two-dimensional (2D) Interaction Diagrams with Dimensionless Parameters

The axial force and moments are normalized based on cross-sectional dimensions ($b \cdot h$) and concrete characteristic strength (f_{ck}). This ensures the diagrams remain applicable regardless of specific dimensions or material strength.

Two-dimensional (2D) interaction diagrams with dimensionless parameters are constructed using normalized values, where the axial load (N) is divided by '$b \cdot h \cdot f_{ck}$', and the total resultant resistive moment (M_{tr}) is divided by '$b \cdot h^2 \cdot f_{ck}$'. If the axial force is expressed in newtons (N), it should be divided by '$b \cdot h \cdot f_{ck}$', while if it is given in kilonewtons (kN), it should be divided by '$b \cdot h \cdot f_{ck} \cdot 10^3$'). Similarly, if the moment (M) is expressed in kilonewton-meters (kNm), it should be divided by '$b \cdot h^2 \cdot f_{ck} \cdot 10^6$' to maintain consistency in units. However, both approaches result in the same dimensionless values.

When the inclination angle (θ) of the neutral axis is set to 0°, the moment resistance aligns with the x-axis, defining the x-axis diagram represented by '$M_x/(bh^2 \cdot f_{ck})$'. Conversely, when the neutral axis is set to 90°, it defines the y-axis diagram represented by '$M_y/(bh^2 \cdot f_{ck})$'. This is shown in Figure 8.8 and Figure 8.9. Moreover, the diagram can also be constructed for custom angles, for instance, the Figure 8.10 shows the interaction diagram for an inclination angle (θ) of 39.805571092°.

The reduction factor ($\omega = 0.9$) used in previous sections does not apply to 90° and 0° inclination angles. This factor accounts for the nonlinear distribution of the compression zone when it decreases toward the extreme compression fibre. Eurocode 2, Clause 3.1.7 (3) states this condition. However, in this case, the neutral axis is set parallel to the x-axis or y-axis. As a result, the extreme compression fibre maintains its full breadth (b) or height (h). Therefore, the clause is not relevant in this context. Consequently, when calculating the moments and forces to construct the 90° and 0° interaction diagrams, the reduction factor (ω) of 0.9 should not be applied. However, if any other angle besides 0° or 90° is used, the 0.9 reduction factor (ω) should be applied.

The data are obtained by shifting the neutral axis forward or backward. The neutral axis starts at the extreme compression fibre intersection point (ECFIP), and it is then shifted gradually until the total resultant resistive moment (M_{tr}) becomes zero or almost zero.

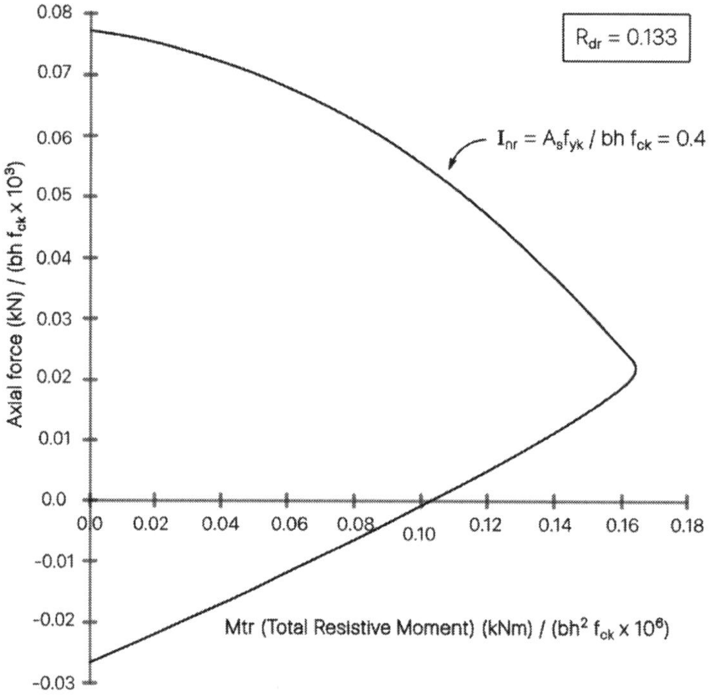

Figure 8.8: Normalized N-M Interaction diagram for x axis at $\theta = 0°$, $\omega = 1.0$ and $R_{dr} = 0.133$

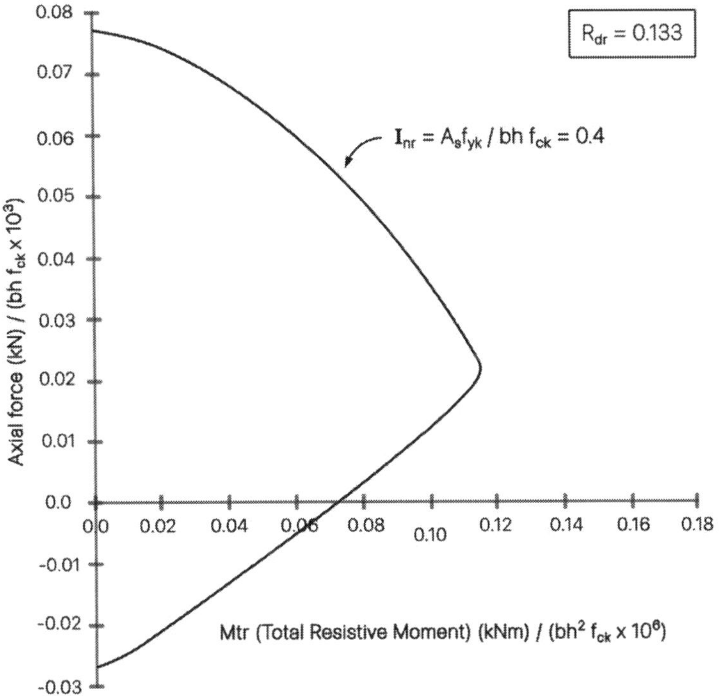

Figure 8.9: Normalized N-M Interaction diagram for y axis at $\theta = 90°$, $\omega = 1.0$ and $R_{dr} = 0.133$

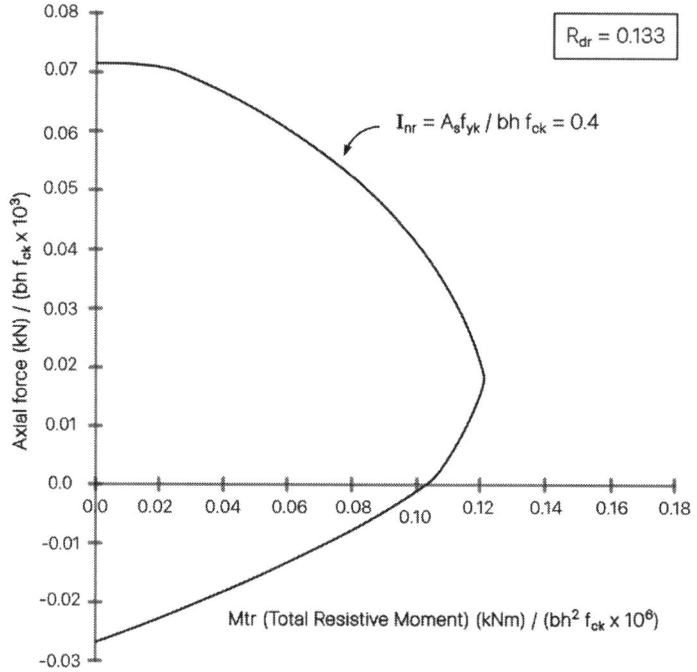

Figure 8.10: Normalized N-M Interaction diagram for y axis at $\theta = 39.805571092°$, $\omega = 0.9$ and $R_{dr} = 0.133$

Chapter 9:

Plotting the Three-dimensional (3D) Interaction Surface

In the previous section, the 2D interaction diagram was constructed by keeping the inclination angle of the neutral axis constant. However, to construct the three-dimensional (3D) interaction surface, a denser and more representative dataset is required. This is achieved by incrementally setting the neutral axis inclination angle from $0°$ to $90°$, typically in small steps (e.g., $5°$ or finer). For each fixed angle, the neutral axis is shifted forward (downward) from the extreme compression fibre intersection point (ECFIP), and at each step, the total axial force (N), the magnitude of the resultant resistive bending moment (M_{tr}), and the direction of the resultant resistive moment (θMd) are recorded.

Importantly, the neutral axis does not rotate during this shifting process. The angle is held constant while the position is progressively shifted until the resultant resistive bending moment (M_{tr}) approaches zero. This process is repeated for each angle in the selected range to generate a complete dataset.

Each inclination angle of the neutral axis generates a distinct 2D interaction curve, analogous to radial profile curves that resemble meridian slices stretching from the South Pole to the North Pole. For each of these angles, the corresponding axial force (N), the total resultant resistive bending moment (M_{tr}), and the direction of the resultant resistive bending moment (θMd) are recorded. These curves, each representing the response of the column at a particular neutral axis orientation, are then connected across the full range of angles to form the complete 3D interaction surface.

Moreover, the 0° and 90° inclination angles correspond to pure uniaxial bending, and therefore do not require the application of the 0.9 reduction factor (as stated in Eurocode 2, Clause 3.1.7 (3)). However, real-world columns are almost always subjected to some degree of biaxial bending. As a result, it is highly unrealistic to expect a biaxially connected and loaded column to behave in a purely uniaxial manner, even in a hypothetical case where the loads themselves are uniaxial. To reflect this reality and maintain consistency, the 0.9 reduction factor has been applied uniformly across all inclination angles, including the theoretical 0° and 90° cases, when generating the data for the 3D N-M interaction surface. This ensures that the surface accurately captures the practical biaxial resistance of columns under combined axial and moment loading.

The surface is not defined using fixed M_x or M_y axes. Instead, it is constructed from the directional nature of the resultant moment vector, where the resultant resistive bending moment (M_{tr}) and its direction together with axial force (N) define the spatial distribution of points. This approach allows for a continuous and smooth representation of the biaxial resistance of the column. This captures a broader and more realistic range of interaction states.

The curves arranged in 3D space, along with different views of the resulting 3D surface, are shown in Figures 9.1 to 9.8. Although these models utilize only a limited number of curves, a smoother surface can be obtained by increasing the number of curves and using smaller angle variations between them.
Creating the interaction surface with smaller increments for neutral axis shifts, that is, adding more data points, reduces the inherent conservatism of the curves and yields interaction diagrams that are correspondingly less conservative.

9.1 Wireframe views of 3D Interaction surface

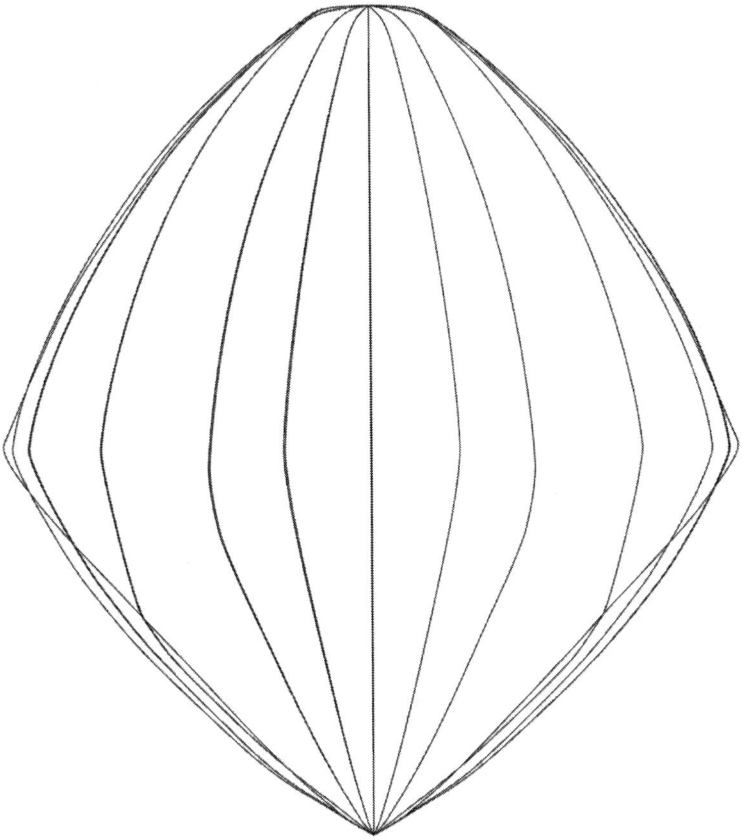

Figure 9.1: Wireframe view of the 3D curve arrangement along y axis

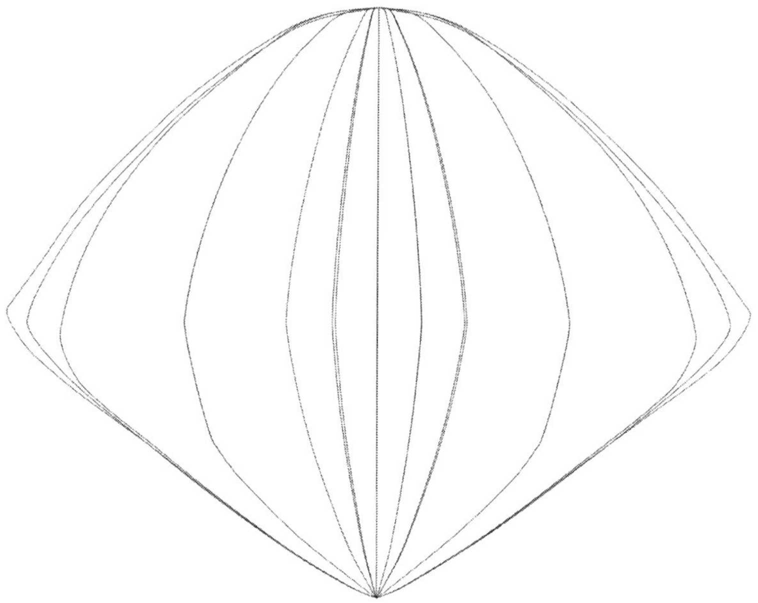

Figure 9.2: Wireframe view of the 3D curve arrangement along x axis

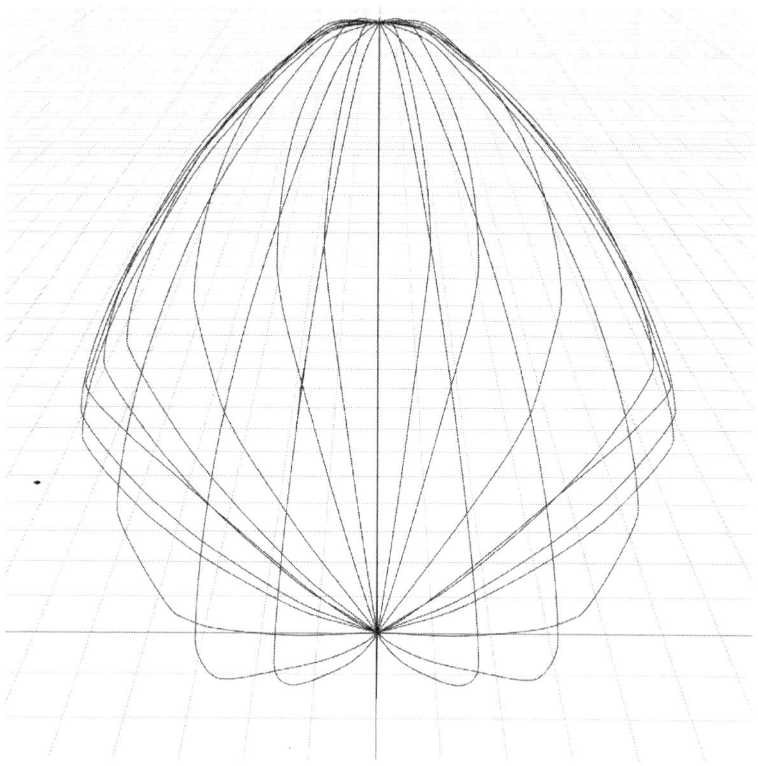

Figure 9.3: Perspective view of the 3D curve arrangement

9.2 Rendered views of 3D Interaction Surface

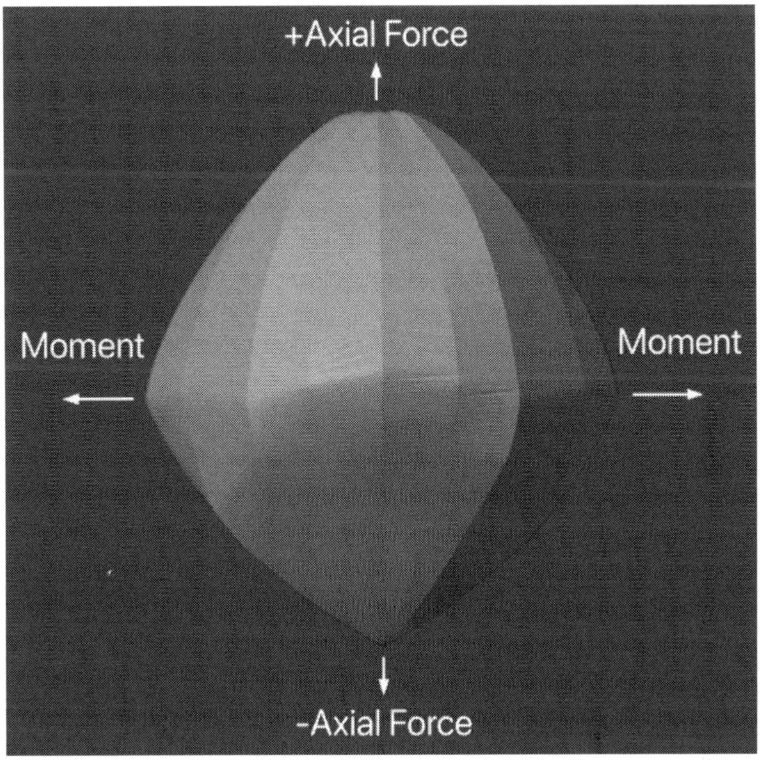

Figure 9.4: 3D view of the N-M interaction surface along the y axis

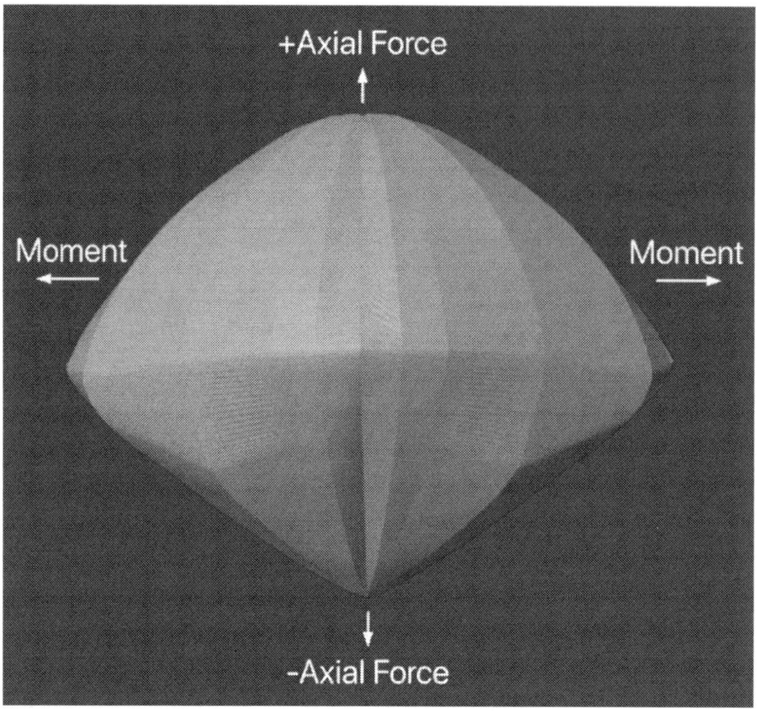

Figure 9.5: 3D view of the N-M interaction surface along the x axis

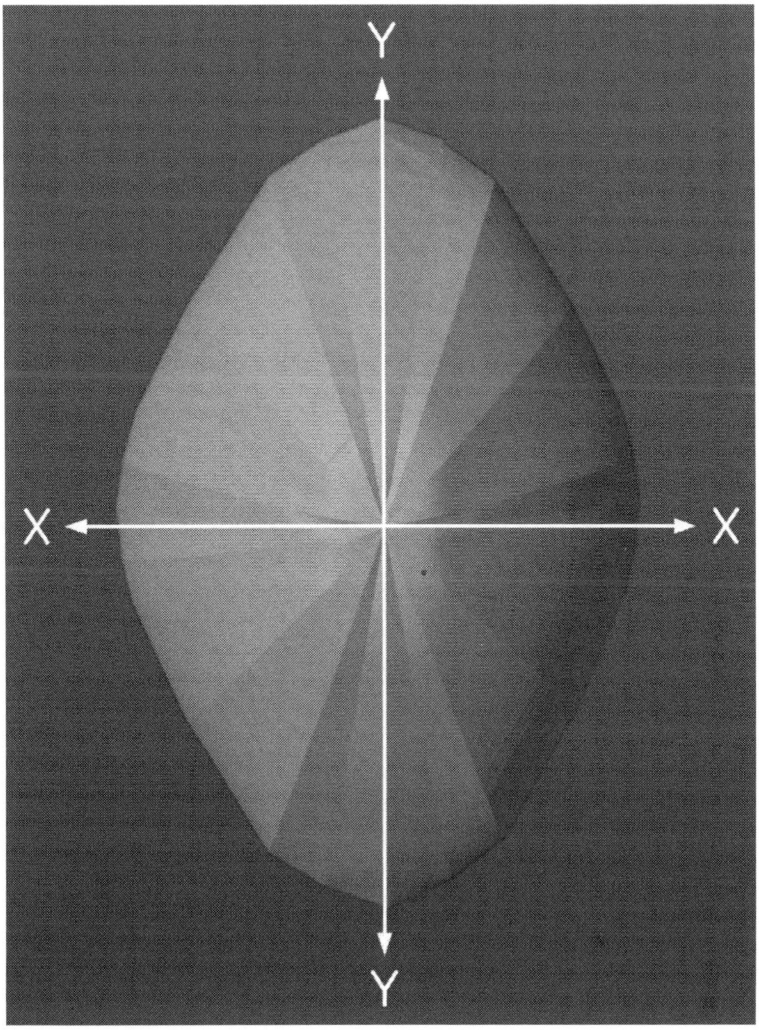

Figure 9.6: Top view of the 3D N-M interaction surface

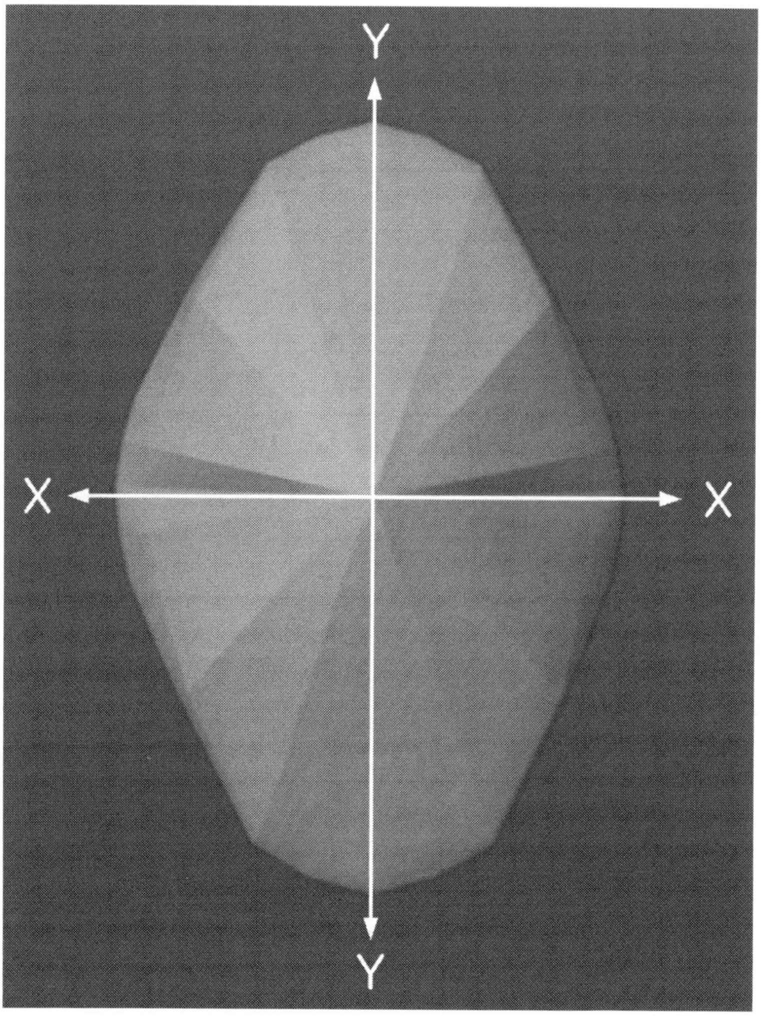

Figure 9.7: Bottom view of the 3D N-M interaction surface

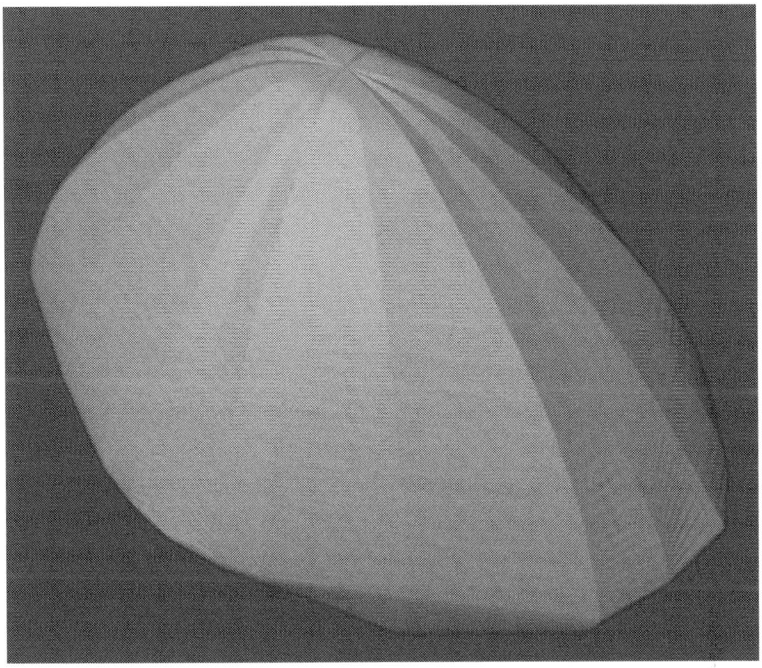

Figure 9.8: Perspective view of the 3D N-M interaction surface

This page intentionally left blank.

Appendix

The data of the diagrams, surface and other information is shown as follows.

A - Data of 2D Interaction Diagrams

Table 9.1: N-M Interaction diagram at $\theta = 0°$ (for x axis), $\omega = 1.0$

Axial Force (kN)	**Total resistive moment (M_{tr}) (kNm)**
-546.364	0.000
-523.153	3.411
-499.193	6.785
-475.233	10.008
-302.733	29.422
-165.881	44.763
-77.340	54.640
0.378	63.176
64.657	70.084
119.975	75.857
169.023	80.789
213.509	85.070
254.574	88.824
293.007	92.139
324.007	94.486
347.967	95.908
371.927	97.180
395.887	98.303
419.847	99.275
443.806	100.097
467.766	100.769
508.750	99.418
564.346	96.311
617.190	93.355
667.627	90.514
715.946	87.756
762.392	85.054
807.172	82.386
850.464	79.730
892.423	77.071
933.183	74.395
972.858	71.687
1011.552	68.938
1049.352	66.136
1075.656	64.448
1111.898	61.518
1147.458	58.513
1182.390	55.426
1216.746	52.254
1250.567	48.989
1283.896	45.629
1316.768	42.169
1349.215	38.605
1381.268	34.935
1412.952	31.156
1444.294	27.264
1475.314	23.257
1506.034	19.133
1565.960	0.000

Table 9.2: N-M Interaction diagram at $\theta = 90°$ (for y axis), $\omega = 1.0$

Axial Force (kN)	**Total resistive moment (M_{tr}) (kNm)**
-546.364	0.000
-542.285	0.457
-517.397	3.176
-488.429	6.189
-459.461	9.036
-336.961	18.500
-184.099	30.001
-83.216	37.569
4.750	44.036
77.966	49.269
141.349	53.624
197.849	57.316
249.343	60.480
297.083	63.208
341.935	65.561
384.518	67.587
423.838	69.210
452.806	69.918
520.650	67.644
587.070	65.308
649.547	63.094
708.673	60.957
764.927	58.865
818.701	56.789
870.318	54.703
920.047	52.590
968.115	50.433
1014.714	48.218
1049.326	46.707
1093.453	44.342
1136.534	41.889
1178.675	39.339
1219.966	36.686
1260.486	33.924
1300.307	31.049
1339.489	28.056
1378.087	24.940
1416.151	21.698
1453.722	18.327
1490.841	14.825
1527.542	11.189
1565.960	0.000

Table 9.3: N-M Interaction diagram at $\theta = 39.805571092°$, $\omega = 0.9$

Axial Force (kN)	**Total resistive moment (M_{tr}) (kNm)**
-546.364	0.000
-545.446	0.167
-542.692	0.651
-538.102	1.424
-497.711	6.932
-424.941	16.697
-377.179	23.132
-334.527	28.831
-297.719	33.670
-264.194	37.959
-217.510	42.700
-155.450	48.289
-98.683	53.530
-45.566	58.419
5.078	62.987
66.573	66.247
128.115	67.721
189.941	69.279
250.099	70.801
309.085	72.243
366.080	73.459
430.457	73.165
499.217	71.776
564.992	70.269
622.813	69.040
683.677	67.219
742.506	65.203
799.528	62.989
854.936	60.553
908.896	57.902
961.554	55.013
1012.118	51.983
1059.780	48.915
1104.637	45.829
1141.433	43.429
1180.927	40.358
1217.843	37.320
1252.243	34.340
1284.181	31.429
1313.703	28.621
1340.855	25.943
1365.670	23.410
1386.209	20.893
1404.576	18.583
1420.794	16.486
1434.884	14.644
1446.864	13.055
1456.752	11.766
1464.564	10.778
1451.210	0.000

B - Supplementary Uniaxial Design Diagrams

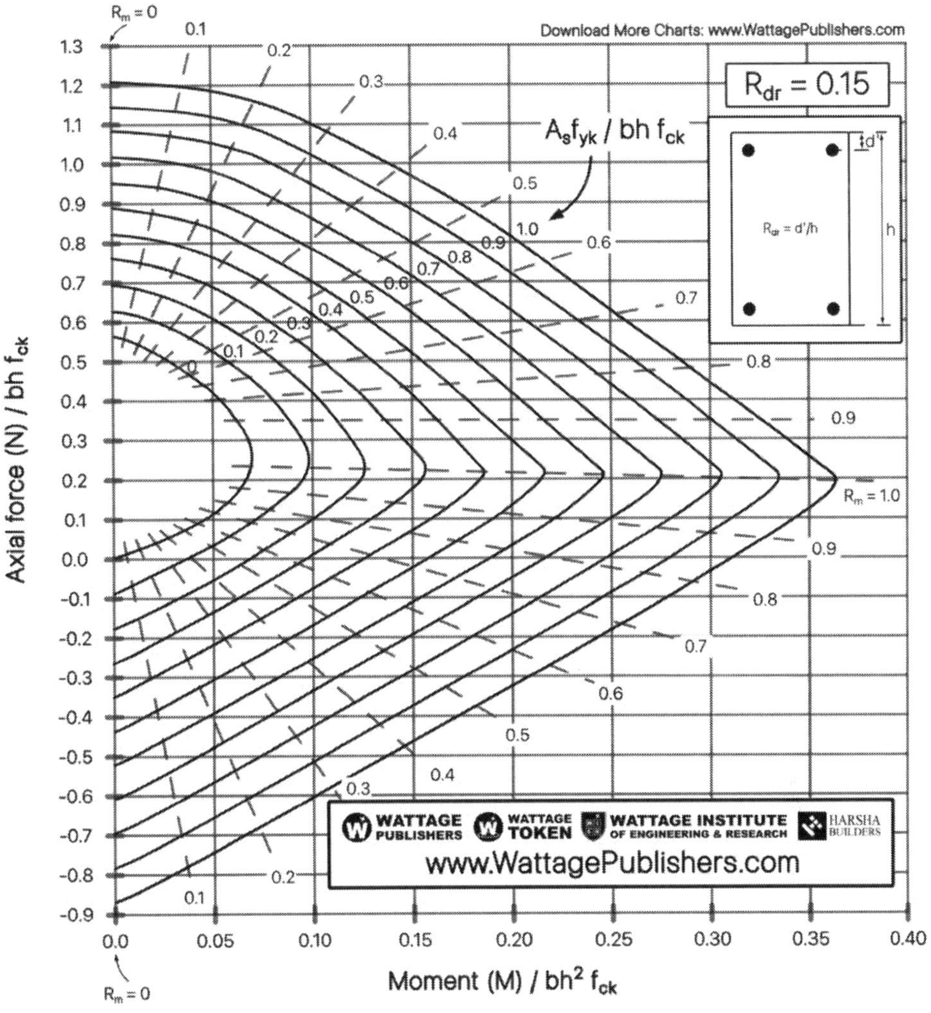

Figure 9.9: Supplementary Uniaxial Design Diagram for x axis at R_{dr} = 0.15

Made in the USA
Monee, IL
12 November 2026

50e525bf-027b-42b0-a975-af260f9c150bR01